T0343994

# GEOMETRY, DYNAMICS AND TOPOLOGY OF FOLIATIONS
A First Course

# GEOMETRY, DYNAMICS AND TOPOLOGY OF FOLIATIONS
A First Course

Bruno Scardua
*Federal University of Rio de Janeiro, Brazil*

Carlos Arnoldo Morales Rojas
*Federal University of Rio de Janeiro, Brazil*

World Scientific

NEW JERSEY · LONDON · SINGAPORE · BEIJING · SHANGHAI · HONG KONG · TAIPEI · CHENNAI · TOKYO

*Published by*

World Scientific Publishing Co. Pte. Ltd.

5 Toh Tuck Link, Singapore 596224

*USA office:* 27 Warren Street, Suite 401-402, Hackensack, NJ 07601

*UK office:* 57 Shelton Street, Covent Garden, London WC2H 9HE

**Library of Congress Cataloging-in-Publication Data**
Names: Scardua, Bruno. | Rojas, Carlos Arnoldo Morales.
Title: Geometry, dynamics, and topology of foliations : a first course / by
    Bruno Scardua (University of Rio de Janeiro, Brazil),
    Carlos Arnoldo Morales Rojas (Federal University of Rio de Janeiro, Brazil).
Description: New Jersey : World Scientific, 2017. | Includes bibliographical references and index.
Identifiers: LCCN 2016059937 | ISBN 9789813207073 (hardcover : alk. paper)
Subjects: LCSH: Foliations (Mathematics) | Differential topology.
Classification: LCC QA613.62 .S33 2017 | DDC 514/.72--dc23
LC record available at https://lccn.loc.gov/2016059937

**British Library Cataloguing-in-Publication Data**
A catalogue record for this book is available from the British Library.

Printed in Singapore

Dedicated to our families

# Preface

The Geometrical Theory of Foliations is one of the fields in Mathematics that gathers several distinctive domains such as; Topology, Dynamical Systems, Differential Topology and Geometry, etc. It originated from the works of C. Ehresmann and G. Reeb ([Ehresmann (1947)]), ([Ehresmann and Reeb (1944)]). The huge development has allowed a better comprehension of several phenomena of mathematical and physical nature. Classical theorems, like the Reeb stability theorem, Haefliger's theorem, and Novikov's compact leaf theorem, are now searched for holomorphic foliations. Several authors have began to investigate such phenomena (e.g. C. Camacho, A. Lins Neto, E. Ghys, M. Brunella, R. Moussu, S. Novikov and others). The study of such field presumes knowledge of results, techniques of the real case and superior familiarity with the classical aspects of Holomorphic Dynamical Systems.

There is a number of important books dedicated to the study of foliations, specially in the non-singular smooth framework. We shall not list all of them, but we cannot avoid mentioning the books of Camacho and Lins Neto ([Camacho and Lins-Neto (1985)]), Candel and Conlon ([Candel and Conlon (2000)]), C. Godbillon ([Godbillon (1991)]) and Hector and Hirsch ([Hector and Hirsch (1987)]), which are among our favorite. Each of these has influenced in our text. From the choice of topics, to the path taken in some demonstrations. In our viewpoint, the book of Camacho and Lins Neto is important for its wise choice of topics, and for aiming at the geometry of foliations. The book of Hector and Hirsch has influenced us specially in the interplay between geometry and dynamics of foliations. The book of C. Godbillon is very interesting for the wide range of topics that are covered. The proofs are elegant, usually short, but still precise. Finally, the book of Candel and Conlon, divided into two volumes, is a very detailed introduction to the general theory of foliations. There, one can find a very complete exposition of the important results in the theory of codimension one foliations. Special attention is given to the theory of minimal sets. This material has greatly influenced our exposition here. Their book is therefore a natural complement to any introduction to the theory of foliations, and should used as a reference any further information and courses.

Finally, we would like to mention that our personal interest for foliations has been greatly supported by some amazing works from several authors. Besides the authors mentioned above or in the text itself we wish to mention P. Schweitzer, D. Calegari, P. Molino, S. Fenley, P. Thondeur and T. Tsuboi.

These notes are mainly introductory and only cover part of the basic aspects of the rich theory of foliations. In particular, additional extensive information in some of the results presented here, may be searched in the bibliography we give. We have tried to clarify the geometry of some classical results and provide motivation for further study. Our goal is to highlight this geometrical viewpoint despite some loss (?) of formalism. We hope that this text may be useful to those who appreciate Mathematics. Specially, to the students that are interested in this exquisite and conducive field of Mathematics.

This text is divided into two basic parts. The first part, which corresponds to the first eight chapters, consists of an exposition of classical results in Geometric Theory of (real) Foliations. Special attention is paid to the classical Reeb Stability theorems, Haefliger's theorem and Novikov's compact leaf theorem.

Starting at Chapter 9, the second part contains a robust proof of Plante's Theorem on growth and compact leaves. This is followed by the basic ingredients of the theory of foliation cycles and currents which is developed in Chapter 10. Then in Chapter 11 this is applied in D. Sullivan's homological proof of Novikov's compact leaf theorem. It is based in a mix of topological argumentation and invariant measure theory for foliations. Chapter A is dedicated to some more specialized results concerning the structure of codimension one foliations on closed manifolds. We present the results of Dippolito on the structure of codimension one foliations and semi-stability. Also we present Catwell-Conlon's result on the minimal sets for such foliations.

We present in Chapter 11, D. Sullivan's homological proof of Novikov's compact leat theorem. This appears to be an applicable procedure for complex foliations. We invite the reader to think about it.

In the last part of the book we give an exposition of some important results on the structure of codimension one foliations. We state results of Dippolito and Cantwell-Conlon on the their structure.

We hope the reader will enjoy reading this book as much as we have enjoyed writing it.

*Carlos Morales and Bruno Scárdua*

# Contents

# Chapter 1

# Preliminaries

## 1.1 Definition of foliation

There are essentially three ways to define foliations. Let $M$ be a $m$-dimensional manifold, $m \in \mathbb{N}$. Let $D^k$ be the open unit ball of $\mathbb{R}^k$ where $k \in \mathbb{N}$. Let $0 \leq n \leq m$ be fixed.

**Definition 1.1.** A $C^r$ *foliation* of codimension $m - n$ of $M$ will be a maximal atlas $\mathcal{F} = \{(U_i, X_j)\}_{i \in I}$ of $M$ satisfying the following properties:

(1) $X_i(U_i) = D^n \times D^{m-n}$;
(2) For all $i, j \in I$ the map $X_j \circ (X_i)^{-1} : X_i(U_i \cap U_j) \to X_j(U_i \cap U_j)$ is $C^r$ and has the form

$$X_j \circ (X_i)^{-1}(x, y) = (f_{i,j}(x, y), g_{i,j}(y)).$$

The number $n$ is called *the dimension* of $\mathcal{F}$. A *plaque* of $\mathcal{F}$ is a set $\alpha = X_i^{-1}(\{y = C\})$ for some $C \in \mathbb{R}^{m-n}$. The plaques of $\mathcal{F}$ define a relation $\approx$ in $M$ as follows: If $x, y \in M$ then $x \approx y$ if and only if there is a finite collection of plaques $\alpha_1, \cdots, \alpha_k$ such that $x \in \alpha_1, y \in \alpha_k$ and $\alpha_i \cap \alpha_{i+1} \neq \emptyset$ for all $1 \leq i \leq k - 1$. Clearly $\approx$ is an equivalence and then we can consider the equivalence class $\mathcal{F}_x$ of $\approx$ containing $x \in M$. A *leaf* of $\mathcal{F}$ is defined as an equivalence class $L = \mathcal{F}_x$ of $\approx$ (for some $x \in M$). One can easily prove that every leaf of $\mathcal{F}$ is an immersed submanifold of $M$. We shall see later that a leaf may self-accumulate, and so, the leaves of $\mathcal{F}$ are not embedded in general. Under the viewpoint of the equivalence $\approx$, one can define $\mathcal{F}$ as a partition of $M$ by immersed submanifolds $L$ such that for all $x \in M$ there is a neighborhood $U$ diffeomorphic to $D^{m-n} \times D^n$ such that the leaves of the partition intersect $U$ in the trivial foliation $\{D^n \times y : y \in D^{m-n}\}$ on

1

$D^{m-n} \times D^n$. Thus, we have the following equivalent definition of foliation.

**Definition 1.2.** A $C^r$ foliation of codimension $m - n$ of $M$ is a partition $\mathcal{F}$ of $M$ consisting of disjoint immersed $C^r$ submanifolds $F \subset M$ with the following property: for each point $x \in M$ there is a neighborhood $U$ of $x$, and a $C^r$ diffeomorphism $X : U \to D^n \times D^{m-n}$, such that $\forall y \in D^{m-n}$ $\exists F \in \mathcal{F}$ satisfying

$$X^{-1}(D^n \times y) \subset F.$$

The elements of the partition $\mathcal{F}$ are called the *leaves of $\mathcal{F}$*. The element $\mathcal{F}_x$ of $\mathcal{F}$ containing $x \in M$ is called the *leaf of $\mathcal{F}$ containing $x$*.

**Warning:** Not every decomposition of $M$ into immersed submanifolds with the same dimension is a foliation. Indeed, consider the partition of $\mathbb{R}^2$ depicted in Figure 1.1 (note that the condition for foliation fails at the point $x$).

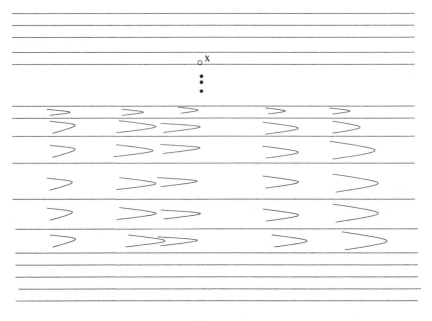

Fig. 1.1

The third definition of foliation uses the notion of distinguished applications. Let $\mathcal{F} = \{(U_i, X_i)\}$ be a foliation of a manifold $M$ in the sense of

**Definition 1.1.** Then $\forall i, j$ the transition map $X_j \circ (X_i)^{-1}$ has the form

$$X_j \circ (X_i)^{-1}(x, y) = (f_{i,j}(x, y), g_{i,j}(y)).$$

The map $g_{i,j}$ is a diffeomorphism in its domain of definition. This follows from the fact that the derivative $D(X_j \circ (X_i)^{-1})(x, y)$ has non-zero determinant equals to $\partial_x f_{i,j}(x, y) \cdot g'_{i,j}(y)$. We define for all $i$ the map $g_i = \Pi_2 \circ X_i$, where $\Pi_2$ is the projection onto the second coordinate $(x, y) \in D^n \times D^{m-n} \to y$. One has $g_j = g_{i,j} \circ g_i$ as $(\Pi_2 \circ X_j) \circ X_i^{-1} = g_{i,j}$ and then $g_{i,j}^{-1} \circ g_j = X_i \Rightarrow \Pi_2 \circ g_{i,j} \circ g_j = \Pi_2 \circ X_i = g_i \Rightarrow g_j = g_{i,j} \circ g_i$ since $\Pi_2$ is the identity in $D^{m-n}$. Therefore, a $C^r$ foliation $\mathcal{F}$ of codimension $m - n$ of a manifold $M^m$ is equipped with a cover $\{U_i\}$ of $M$ and $C^r$ submersions $g_i : U_i \to D^{m-n}$ such that for all $i, j$ there is a diffeomorphism $g_{i,j} : D^{m-n} \to D^{m-n}$ satisfying the cocycle relations

$$g_j = g_{i,j} \circ g_i, \quad g_{i,i} = \mathrm{Id}.$$

The $g_i$'s are the *distinguished maps* of $\mathcal{F}$.

Conversely, suppose that $M^m$ admits an open cover $M = \bigcup_{i \in I} U_i$ such that for each $i \in I$ there is a $C^r$ submersion $g_i : U_i \to D^{m-n}$ such that for all $i, j$ there is a diffeomorphism $g_{i,j}$ satisfying the cocycle relations above. By the Local form of the submersions we can assume that for each $i \in I$ there is a $C^r$ diffeomorphism $X_i : U_i \to D^n \times D^{m-n}$ such that

$$g_i = \Pi_2 \circ X_i$$

since

$$\Pi_2 \circ X_j \circ (X_i)^{-1}) = g_j \circ (X_i)^{-1} = g_{i,j} \circ g_i \circ (X_i)^{-1} = g_{i,j} \circ \Pi_2,$$

we have that the atlas

$$\mathcal{F} = \{(U_i, X_i)\}$$

defines a foliation of class $C^r$ and codimension $m - n$ of $M$. The above suggests the following equivalent definition of foliation.

**Definition 1.3.** A $C^r$ *foliation* of codimension $m - n$ of $M$ is a cover $\{U_i : i \in I\}$ of $M$ such that $\forall i \in I$ there is a $C^r$ submersion $g_i : U_i \to D^{m-n}$ such that $\forall i, j \in I$ there is a diffeomorphism $g_{i,j} : D^{m-n} \to D^{m-n}$ satisfying the cocycle relations

$$g_j = g_{i,j} \circ g_i, \quad g_{i,i} = \mathrm{Id}.$$

The $g_i$'s are the *distinguished applications* of $\mathcal{F}$.

This last definition leads to several interesting definitions. For instance, a foliation $\mathcal{F}$ of $M$ is said to be *transversely orientable* or *transversely affine* depending on whether, for some convenient choice, its distinguished applications $g_{i,j}$ are orientation preserving or affine maps. An equivalent definition will be given in Section 2.2. In order to distinguish foliations, we shall use the following definition.

**Definition 1.4.** Two foliations $\mathcal{F}, \mathcal{F}'$ defined on $M, M'$ respectively are $C^r$-*equivalent* if there is a $C^r$-diffeomorphism $h \colon M \to M'$ ($h$ is a homeomorphism if $r = 0$), sending leaves of $\mathcal{F}$ into leaves of $\mathcal{F}'$. In other words,

$$h(\mathcal{F}_x) = \mathcal{F}'_{h(x)}, \ \forall\, x \in M.$$

The above relation defines an equivalence in the space of foliations. As an illustration, observe that the foliations $F_1, F_2$ in the band $I \times \mathbb{R}$ in Figure 1.2 are not equivalent.

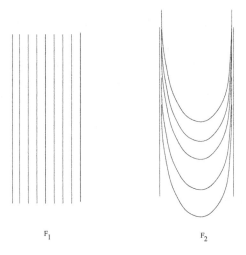

$F_1$                    $F_2$

Fig. 1.2   Non equivalent foliations.

## 1.2    Examples of foliations

### 1.2.1    *Foliations derived from submersions*

A *submersion* between two manifolds is a smooth map whose derivative has maximal rank everywhere. Submersions provide the very first examples of foliations.

**Theorem 1.1.** *Let $f\colon M^m \to N^n$ be a $C^r$ submersion. Then the connected components of the level submanifolds*

$$L_c = f^{-1}(c), \quad c \in N$$

*are the leaves of a $C^r$ foliation of codimension $n$ of $M$.*

**Proof.** By the Local Form of the Submersions [do Carmo (1992)] there are atlases $\{(U, X)\}, \{(V, Y)\}$ of $M, N$ respectively such that

(1) $X(U) = D^n \times D^{m-n}$.
(2) $Y(V) = D^{m-n}$.
(3) $Y \circ f \circ X^{-1} = \Pi_2$ (see Figure 1.3).

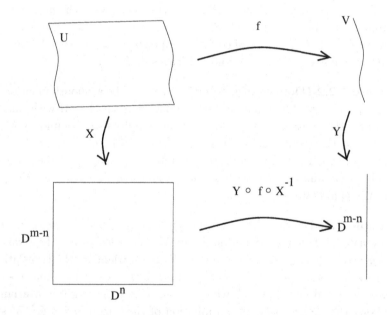

Fig. 1.3   Foliation and submersion.

We claim that the collection $\mathcal{F} = \{(U, X)\}$ defines a foliation of $M$. In fact, let $(U, X), (U^*, X^*)$ be two elements of the cover. Then,

$$\Pi_2 \circ X^* \circ X^{-1} = Y^* \circ f \circ (X^*)^{-1} \circ X^* \circ X^1$$

$$= Y^* \circ f \circ X^{-1} = Y^* \circ Y^{-1} \circ Y \circ f \circ X^{-1} = Y^* \circ Y^{-1} \circ \Pi_2.$$

Hence $\Pi_2 \circ (X^* \circ X^{-1}) = (Y^* \circ Y^{-1}) \circ \Pi_2$ does not depend on $x \in D^n$. This proves that $\mathcal{F}$ is a foliation of class $C^r$ and codimension $m - n$ of $M$. It is clear by definition that the plaques of $\mathcal{F}$ are contained in the level sets of $f$. This proves that the leaves of $\mathcal{F}$ are precisely the level sets of $f$ and the result follows. $\qquad\square$

Let us present some examples to illustrate the above result.

**Example 1.1.** Let $M$ and $N$ be $C^r$ manifolds and take $f \colon M \times N \to M$ as the first coordinate projection $f(x, y) = x$. Clearly $f$ is a $C^r$-submersion. In this case $f$ defines a $C^r$-foliation of $M \times N$ whose leaves are the vertical fibers in $\{x\} \times N$, $x \in M$.

Exercise 1.2.1. Let $\mathcal{F}$ be a foliation on $M$ of codimension $q$. A differentiable map $f \colon N \to M$ is *transverse* to $\mathcal{F}$ if it is transverse to each leaf $L \in \mathcal{F}$ as an immersed submanifold in $M$. Show that in this case there is a naturally defined foliation $f^*(\mathcal{F})$ in $N$ of codimension $q$ such that for each leaf $L \in \mathcal{F}$ the inverse image $f^{-1}(L)$ is a union of leaves of $f^*(\mathcal{F})$.

**Exercise 1.2.2 (Double of a foliation).** Let $\mathcal{F}$ be a smooth foliation on $M$. Suppose that we have a relatively compact domain $D \subset M$ with smooth boundary $\partial D$ transverse to $\mathcal{F}$. Consider the manifold with boundary $M_0 = M \setminus D$ and the restriction $\mathcal{F}_0 = \mathcal{F}\big|_{M_0}$. Given two copies $M_1$ and $M_2$ of $M_0$ we can construct a manifold $M_d$ by gluing these copies by the common boundary $\partial D$ and equip it with a smooth foliation $\mathcal{F}_d$ such that $\mathcal{F}_d\big|_{M_j}$ is naturally conjugate to $\mathcal{F}_0$.

**Example 1.2.** Let $M = \mathbb{R}^2$ and $f(x, y) = y - \alpha \cdot x$, where $\alpha \in \mathbb{R}$. The level curves of $f$ define a foliation $\mathcal{F}_\alpha$ in $M$ whose leaves are the straight-lines $y = \alpha \cdot x + c$, $c \in \mathbb{R}$. Observe that $\mathcal{F}_\alpha$ is invariant by the translations $(x, y) \to (x + k, y + l)$, $(k, l) \in \mathbb{Z}^2$. Indeed, if $y = \alpha \cdot x + c$ then $y + l = \alpha \cdot x + c + l = \alpha \cdot (x + k) + c'$, where $c' = c - \alpha \cdot k$ proving the invariance. It follows that $\mathcal{F}_\alpha$ projects into a foliation of the 2-torus $T^2 = \mathbb{R}^2/\mathbb{Z}^2$ still denoted by $\mathcal{F}_\alpha$. See Figure 1.4. When $\alpha$ is irrational then all the leaves of the induced foliation are lines, and if $\alpha$ is rational then all the leaves are circles. We call this example as the *linear foliation* in $T^2$.

**Example 1.3 (Reeb foliation on $S^3$).** Let $M = \mathbb{R}^3$ and $f(x, y, z) = \alpha(r^2)e^z$, where $r^2 = x^2 + y^2$ and $\alpha$ is a $C^\infty$ function such that $\alpha(0) = 1, \alpha(1) = 0$ and $\alpha'(t) < 0$ for all $t > 0$ (see Figure 1.5).

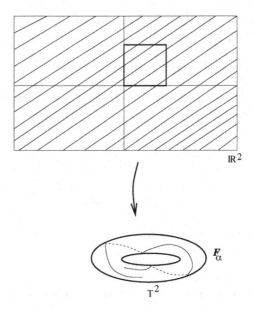

Fig. 1.4   Linear foliation on $T^2$.

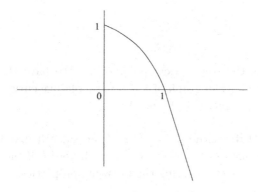

Fig. 1.5   Graph of $\alpha$.

The map $f$ is a submersion since

$$\nabla f(x, y, z) = (2\alpha'(r^2)xe^z, 2\alpha'(r^2)ye^z, \alpha(r^2)e^z) = (0, 0, 0)$$

$\Rightarrow x = y = 0$ and $\alpha(r^2) = 0 \Rightarrow x = y = 0$, $x^2 + y^2 = 1$ contradiction. Hence $\nabla f(x, y, z)$ does not vanish and so $f$ is a submersion. It follows from Theorem 1.1 that the level curves $f^{-1}(c)$ define a foliation of class $C^\infty$ and codimension 1 of $M$. The leaves of this foliation, (*i.e.*, the level curves of $f$) can be described as follows.

$$f(x, y, z) = c \Leftrightarrow \alpha(r^2)e^z = c.$$

If $c = 0$ then $\alpha(r^2) = 0 \Rightarrow x^2 + y^2 = 1$. Hence the level curve corresponding to $c = 0$ is the cylinder $x^2, y^2 = 1$ in $M$. If $c > 0$ then

$$\alpha(r^2)e^z = c \Rightarrow \alpha(r^2) > 0.$$

Moreover,

$$z = K - \ln(\alpha(r^2)),$$

$(K = \ln(c))$. When $c = 1$ we have

$$z = -\ln(\alpha(r^2)).$$

The graph of the above curve in the plane $y = 0$ is given by

$$z = -\ln(\alpha(x^2)).$$

We have

$$z' = -\frac{2\alpha(x^2))}{\alpha(x^2))} \cdot x = 0 \Rightarrow x = 0.$$

Hence $x = 0$ is the sole critical point of $z$. We have that $z \to \infty$ as $x \to x \to 1^+$ or $1^-$. The graph of $z$ is a parabola-like curve. The graph of the leaves of $\mathcal{F}$ is depicted in Figure 1.6.

**Example 1.4 (Fibrations).** Let $E, B, F$ be smooth manifolds. We say that $E$ is a *fiber bundle* of class $C^r$ over $B$ with fiber $F$ if there is a $C^r$ onto submersion $\pi : E \to B$ satisfying the following properties:

(1) $\pi^{-1}(b)$ is diffeomorphic to $F$, $\forall b \in B$.
(2) For each point $b \in B$ there exists a neighborhood $U \subset B$ of $b$ and a diffeomorphism $\phi : \pi^{-1} \to U \times F$ such that $\pi^1 \circ \phi = \pi$, where $\pi^1$ is the projection onto the first coordinate in $U \times F$.

Clearly the family $\{\pi^{-1}(b) : b \in B\}$ is a $C^\infty$ codimension $\dim(B)$ foliation of $E$ since $\pi$ is a $C^\infty$ submersion. Note that the leaves of the resulting foliation are all diffeomorphic to a common manifold $F$.

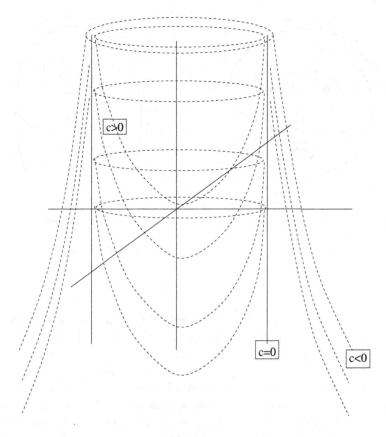

Fig. 1.6

## 1.2.2 *Reeb foliations*

Several are the classes of foliations called *Reeb foliations*. The very first ones
are the Reeb foliations in the cylinder and in the Moebius band, constructed
as follows. Define $M = [-1, 1] \times \mathbb{R}$ and let $\mathcal{F}$ be the foliation in $M$ defined by
the submersion $g(x, y) = \alpha(x^2)e^y$ where $\alpha$ is decreasing. Let $G \colon M \to M$ be
given by $G(x, y) = (x, y + 2)$. The quotient manifold $M/G$ is the cylinder.
Analogously we can replace $G$ by the map $F(x, y) = (-x, y + 2)$. In this
case the quotient manifold $M/F$ is the Moebius band. In each case one
can see that $\mathcal{F}$ is invariant for $G$ and $F$. Hence $\mathcal{F}$ induces a foliation $\overline{\mathcal{F}}$ in
either $M/G$ or $M/F$. These are the Reeb foliations in the cylinder and the
Moebius band respectively. These foliations are depicted in Figure 1.7.

 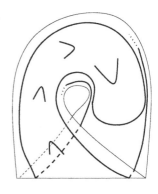

Reeb foliation in the cylinder　　　　　Reeb foliation in the Moebious band

Fig. 1.7　Reeb foliation in the cylinder and Moebius band.

Consider the foliation $\mathcal{F}$ constructed in Example 1.3 of Section 1.2 restricted to the solid cylinder $\{(x, y, z) : x^2 + y^2 \leq 1\}$. One can easily check that the leaves of this foliation are invariant by the translations $(x, y, z) \mapsto (x, y, z + 1)$. Note that the quotient manifold solid cylinder$/(x, y, z) \to (x, y, z + 1)$ is a solid torus $D^2 \times S^1$. The invariance mentioned above implies that $\mathcal{F}'$ induces a foliation in $D^2 \times S^1$ whose leaves are depicted in Figure 1.8. This foliation is called the *Reeb foliation* in the solid torus $ST = D^2 \times S^1$. The Reeb foliation in the solid torus is used to construct a $C^\infty$ foliation in the 3-sphere $S^3$ in the following way:

Let $ST_1$ and $ST_2$ be two solid tori and denote $\partial ST_1 = T_1$ and $\partial ST_2 = T_2$ the corresponding boundaries. Consider a diffeomorphism $\varphi \colon T_2 \to T_1$ sending the meridian curves in $T_2$ into the parallel curves in $T_1$. For instance we can choose $\varphi$ by first considering

$$\varphi(x, y) = \begin{pmatrix} 0 & 1 \\ 1 & 0 \end{pmatrix} \cdot \begin{pmatrix} x \\ y \end{pmatrix}.$$

Because $\varphi$ is linear and $\det \varphi = -1$ we have $\varphi(\mathbb{Z}^2) = \mathbb{Z}^2$ and then $\varphi$ defines the desired map.

In $ST_1 \cup ST_2$ we consider the equivalence relation given by $y = \varphi(x)$. In other words we use the identification below.

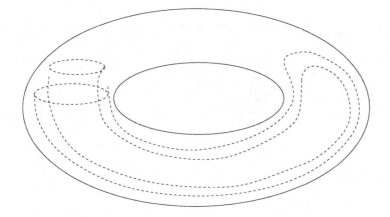

Fig. 1.8   Reeb foliation in the solid torus.

$$x \sim y \Leftrightarrow \begin{cases} x, y \in \text{Int } ST_2 \text{ and } x = y, \text{ or} \\ x \in \text{Int } ST_1, y \in \text{Int } ST_1 \text{ and } x = y, \text{ or} \\ x \in T_1, \ldots, y \in T_2 \text{ and } \varphi(y) = x \end{cases} \qquad (1.1)$$

Consider the quotient manifold $M = (ST_1 \cup ST_2)/\sim = ST_1 \cup_\varphi ST_2$.

**Claim 1.1.** $M = S^3$.

In the proof of this claim we use:

**Alexander's trick** ([Hansen (1989)]): Let $B_1, B_2$ two closed 3-balls, $S_1 = \partial B_1$ and $S_2 = \partial B_2$ be the corresponding 2-sphere boundaries. If $\varphi \colon S_1 \to S_2$ is a diffeomorphism then $B_1 \cup_\varphi B_2 = S^3$.

Now we return to the proof of the claim. Take a region $P$ in between two meridians of $ST_1$. Delete $P$ and cap it into the torus hole in $ST_2$ as explained in Figure 1.9. With this procedure we obtain two 3-balls whose union along the corresponding boundaries yields $S^3$ by the Alexander trick. This proves the claim. The *Reeb foliation* in $S^3$ is precisely the one obtained by the gluing map $\varphi$ setting inside each solid torus the Reed foliation of the solid torus (see Figure 1.10). The above construction leads to construct foliations with only one compact leaf in any manifold of the form $ST_1 \cup_\varphi ST_2$. For instance if $\varphi$ were the identity map then the resulting manifold is $S^2 \times S^1$. We have then constructed a foliation with just one

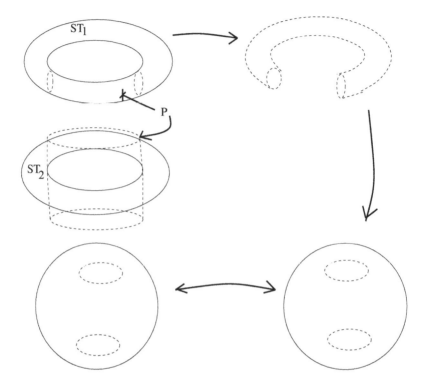

Fig. 1.9

compact leaf in $S^2 \times S^1$. The resulting foliation is clearly different to the one obtained by the trivial fibration $\{S^2 \times y : y \in S^1\}$ of $S^2 \times S^1$.

Exercise 1.2.3. Show that the Reeb foliation in $S^3$ cannot be obtained from a submersion.

**Exercise 1.2.4 (Novikov).** Show that a vector field transverse to the Reeb foliation in $S^3$ has a periodic orbit. Observe that there are non-singular $C^\infty$ vector fields in $S^3$ without periodic orbits (these vector fields are precisely the counterexamples for the Seifert Conjecture [Seifert (1950); Schweitzer (1974)]). Use Novikov's compact leaf theorem (see Chapter 6) to show that any vector field transverse to a codimension one $C^2$ foliation in a compact manifold with finite fundamental group has a periodic orbit.

**Definition 1.5 (Reeb component).** A *Reeb component* of a codimension one foliation $\mathcal{F}$ in $M^3$ is a solid torus $ST \subset M^3$ which is union of leaves of $\mathcal{F}$ such that $\mathcal{F}$ restricted to $ST$ is equivalent to the Reeb foliation in the solid torus $D^2 \times S^1$. A foliation is said to be *Reebless* if it has no Reeb components. Novikov's compact leaf theorem in Chapter 6 states that a compact 3-manifold supporting a Reebless foliation must have an infinite fundamental group.

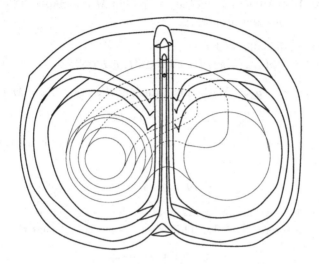

Fig. 1.10   Reeb foliation in $S^3$.

### 1.2.3   *Lie group actions*

A *Lie group* is a group $(G, \cdot)$ with a differentiable structure making the maps

$$\begin{array}{ccc} G \times G \to & G \\ x, y \mapsto x \cdot y \end{array} \quad \text{and} \quad \begin{array}{ccc} G \to & G \\ x \mapsto x^{-1} \end{array}$$

differentiable.

**Example 1.5.** $(\mathbb{R}^n, +)$ is a Lie group, therefore $S^n \setminus \{\text{point}\}$ is a Lie group via stereographic projection. If $\mathbb{C}$ denotes the set of complex numbers (with the complex product), then $\mathbb{C}^* = \mathbb{C} - \{0\}$ with the complex number product is a Lie group. The circle $S^1 \subset \mathbb{C}^*$ is a Lie group when equipped with the product induced by $\mathbb{C}$. Actually it is a Lie subgroup of $\mathbb{C}^*$. The $n$-torus $T^n = \underbrace{S^1 \times \cdots \times S^1}_{n \text{ copies}}$ with the product $(z_1, \ldots, z_m)(z_1^1, \ldots, z_m^1) =$

$(z_1 \cdot z_1^1, \ldots, z_n \cdot z_m^1)$, $z_i, z_i^1 \in S^1$ is a compact Lie group.[1]    The linear group $\mathrm{GL}_n(\mathbb{R}) = \{A \in M_{n \times n}(\mathbb{R}), \det A \neq 0\}$ with the usual matrix product is a Lie group. The subgroup $\mathcal{O}(n) \subset \mathrm{GL}_n(\mathbb{R})$ of orthonormal matrices is a compact Lie subgroup of $\mathrm{GL}_n(\mathbb{R})$.

**Definition 1.6.** An *action* of a Lie group $G$ in $M$ is a map $\varphi : G \times M \to M$ satisfying the following properties:

(1) $\varphi(e, x) = x$ for all $x \in M$;
(2) $\varphi(g \cdot h, x) = \varphi(g, \varphi(h, x))$ for all $x \in M$, $g, h \in G$.

We write $\varphi(g, x) = g \cdot x$. With this notation, the *orbit* of $x \in M$ is the set

$$O_x = \{g \cdot x : g \in G\}.$$

The *isotropy group* of $x \in M$ is the subgroup $G_x \subset G$ consisting of the elements $g \in G$ fixing $x$, namely

$$G_x = \{g \in G : g \cdot x = x\}.$$

**Definition 1.7.** An action $\varphi : G \times M \to M$ is *locally free* if the isotropy group $G_x$ is discrete, $\forall x \in M$. This is equivalent to say that the map

$$\varphi_x : g \in G \mapsto g \cdot x,$$

is an immersion for all $x \in M$ fixed.

The action $G \times M \to M$ is locally free if, and only if, the orbit $O_x$ of $x$ is an immersed submanifold of $M$ with constant dimension $\dim(O_x) = \dim(G)$. Since $G_x < G$ is a closed subgroup it is itself a Lie group (Cartan's Theorem) and also the quotient $G/G_x$ has the structure of a differentiable manifold. Actually we have $G_x = G_y \; \forall x, y$ belonging to some orbit of $p$ and we may introduce the *isotropy subgroup of an orbit* as well.

Given any $x \in M$ we have a natural (diffeomorphism) identification $G/G_x \cong \mathcal{O}_x$. This gives an immersed submanifold structure $\mathcal{O}_x \hookrightarrow M$.

**Theorem 1.2.** *The orbits of a $C^r$ locally free action of a Lie group $G$ on a manifold $M^m$ are the leaves of a $C^r$ foliation of codimension $m - \dim(G)$ of $M$.*

**Proof.** Let $\varphi : G \times M^m \to M^m$ be a locally free $C^r$ action. Fix a point $z_0 \in M$ and set $n = \dim(G)$. By assumption $\dim(O_{z_0}) = n$.

---

[1]The product of Lie groups is a Lie group with a natural product structure ([Hirsch (1971)]).

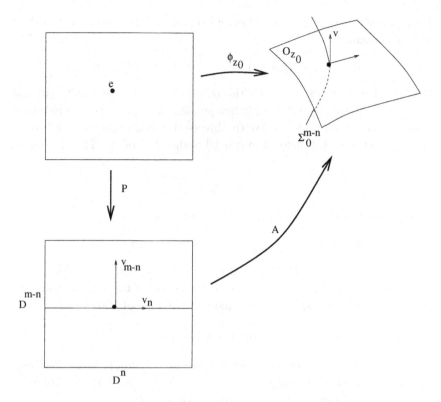

Fig. 1.11  Locally free action.

Let $\Sigma_0^{m-n}$ be a $m - n$ dimensional submanifold transverse to $O_{z_0}$ at $z_0$. Let $(P, U)$ be a local chart at the unity $e \in G$ such that $P(U) = D^n$. Let $B : (D^{m-n}, 0) \to (\Sigma_O^{m-n}, z_0)$ be a parametrization of $\Sigma_0^{m-n}$. Recalling the notation $\varphi(g, x) = g \cdot x$ we define the map $A : D^n \times D^{m-n} \to M$ by

$$A(x, y) = P^{-1}(x) \cdot B(y).$$

The derivative $DA(x, y)$ is given by the expression below:

$$\partial_g \varphi(P^{-1}(x), B(y)) \cdot DP^{-1}(x) + \partial_z \varphi(P^{-1}(x), B(y)) \cdot DB(y).$$

Replacing by $(x, y) = (0, 0)$ one has

$$DA(0, 0) = D\varphi_{z_0}(e) \cdot DP^{-1}(0) + \partial_z \varphi(e, z_0) \cdot B(0)$$

$$= D\varphi_{z_0}(e) \cdot DP^{-1}(0) + DB(0).$$

Let us write $v = v_n \oplus v_{m-n}$ for a tangent vector $v$ of the product $D^n \times D^{m-n}$ at $(0,0)$. Hence

$$DA(0,0) \cdot v = D\varphi_{z_0}(e) \cdot DP^{-1}(0) \cdot v_n + DB(0) \cdot v_{m-n}.$$

Hence $DA(0,0) \neq 0$ if $v \neq 0$ for the vectors $v'_n = D\varphi_{z_0}(e) \cdot DP^{-1}(0)$ and $v'_{m-n} = DB(0) \cdot v_{m-n}$ are linearly independent in $T_{z_0}M$. We conclude that $A$ is a local diffeomorphism. By the Inverse Function theorem the inverse $X = A^{-1}$ of $A$ is well defined in a neighborhood $U$ of $z_0$. This defines an atlas

$$\mathcal{F} = \{(X, U)\}$$

of $M$. Note that for all fixed $y_0 \in D^{m-n}$ one has

$$A(\{(x, y_0) : x \in D^n\}) = \{P^{-1}(x) \cdot B(y_0) : x \in D^n\}.$$

Hence $A(\{(x, y_0) : x \in D^n\})$ is contained in the orbit $O_{B(y_0)}$. This proves that the plaques of $\mathcal{F}$ are contained in the orbits of the action. Since the orbits are pairwise disjoint we conclude that $\mathcal{F}$ is a foliation of $M$. $\quad\square$

In a similar way one can prove the following.

**Theorem 1.3.** *The orbits $O_x$ of a $C^r$ action on a manifold $M$ are the leaves of a $C^r$ foliation if and only if the map $M \ni x \mapsto \dim(O_x)$ is constant. In this case the action is called a* foliated action *on $M$.*

**Example 1.6.** If $G$ is a Lie group and $H$ is a Lie subgroup of $G$ then there is a natural action $H \times G \to G$ by left multiplication. This action is foliated since the left translation is a diffeomorphism and so the orbits of the action define a foliation of $G$. Note that the action is locally free if, and only if, $H$ is discrete.

**Example 1.7.** A $C^r$ *flow* on a manifold $M$ is an action $\varphi$ of the additive Lie group $\mathbb{R}$ in $M$. Denote by $X$ the corresponding tangent vector field $X(x) := \frac{\partial}{\partial t}\big|_{t=0}\varphi(t, x)$. Then $X$ is non-singular if and only if $\varphi$ is locally free. The orbits of $\varphi$ are either circles or lines. In the first case the orbit is called *periodic* and the flow is *periodic* if all its orbits are periodic. A manifold is called *Seifert* if it admits periodic flows.

Exercise 1.2.5. Let $G < \mathbb{R}^2$ be a discrete subgroup. Show that $G$ is isomorphic to $\mathbb{Z}$ or to $\mathbb{Z}^2$. Conclude that the orbits of a locally free action of the affine group $\mathrm{Aff}(\mathbb{R})$ are either planes or cylinders.

**Theorem 1.4 (Hopf's theorem, [Hopf (1931)]).** *The three-sphere $S^3$ is a Seifert manifold.*

Indeed, note that $S^3 = \{(z_1, z_2) \subset \mathbb{R}^2 \times \mathbb{R}^2 : ||z_1||^2 + ||z_2||^2 = 1\}$. Define $Q \colon S^1 \times S^3 \to S^3$ by $Q(x, (z_1, z_2)) = (x \cdot z_1, x \cdot z_2)$. One sees that $Q$ is an action of $S^1$ in $S^3$. The orbits of $Q$ define a foliation by circles in $S^3$ proving that $S^3$ is Seifert. The resulting flow is called *Hopf fibration* of $S^3$ (see [Lyons (2003)] for more details). Seifert manifolds are important in 3-manifold topology since the ones with infinite fundamental group can be described by their fundamental groups. This fact was discovered by Scott (see [Scott (1983)] for a very complete text on the subject). The fundamental group classification of Seifert manifolds with finite fundamental group is false because there are homotopy equivalent lens spaces which are not homeomorphic.

### 1.2.4 $\mathbb{R}^n$ *actions*

We shall describe the actions of the additive group $\mathbb{R}^n$ on a manifold $M$. Let $Q \colon \mathbb{R}^n \times M \to M$ an action and $\{e_1, \ldots, e_n\}$ be a basis of $\mathbb{R}^n$. Fix $i = 1, \ldots, n$ and consider a map

$$\mathbb{R} \times M \xrightarrow{X^i} M$$
$$(t, x) \mapsto Q(te_i, x)$$

This map defines an action of $\mathbb{R}$ on $M$. In fact $X^i(t, x) = Q(t \cdot e_i, x)$ is a flow in $M$. We still denote by $X^i$ the vector field induced by $X^i$, namely

$$X^i(X_t^i(x)) = \frac{d}{dt}(X_t^i(x)) \quad \text{and} \quad X^i(x) = \frac{d}{dt}\bigg|_{t=0} Q(t \cdot e_i, x).$$

In this way we have $n$ vector fields $X^1, \ldots, X^n$ in $M$ such that $X_t^i(x) = Q(t \cdot e_i, x), \quad \forall t \in \mathbb{R}$.

Now, let $v = \sum_{i=1}^{n} t_i e_i \in \mathbb{R}^n$ and $x \in M$ be fixed. Then

$$Q(v, x) = Q\left(\sum_{n=1}^{n} t_i e_i, x\right) = Q\left(\sum_{i=1}^{n-1} t_i e_i, Q(t_n e_m, x)\right)$$

$$= Q\left(\sum_{i=1}^{n-1} t_i e_i, X_{t_n}^n(x)\right) = X_{t_1}^1 \circ X_{t_2}^2 \circ \cdots \circ X_{t_n}^n(x), \forall x \in M.$$

Hence, for all action $Q \colon \mathbb{R}^n \times M \to M$, there are vector fields $X^1, \ldots, X^m$ on $M$ such that $Q(v, x) = X^1_{t_1(v)} \circ \cdots \circ X^n_{t_n(v)}(x)$, $\forall\, x \in M$, $\forall\, v \in \mathbb{R}^n$.

Note that $X^1, \ldots, X^n$ pairwise commute, namely $X^i_t \circ X^j_s = X^j_x \circ X^i_t$. In fact, we have $(X^i_t \circ X^j_s)(x) = Q(t e_i, X^j_s(x)) = Q(t\, e_i, Q(s\, e_j, x)) = Q(t\, e_i + s\, e_j, x) = Q(s\, e_j + t\, e_i, x) = X^j_s \circ X^i_t(x)$.

It is known that $X$, $Y$ commute $\Leftrightarrow [X, Y] = 0$.

Conversely, if $X^1, \ldots, X^n$ are pairwise commuting vector fields, then they define an action $Q$ given by

$$Q(v, x) := X_{t_1(v)}(x) \circ \cdots \circ X_{t_n(v)}(x).$$

**Definition 1.8.** The *rank* of a closed manifold $M$ is the maximal number of linearly independent pairwise commuting vector fields defined in $M$.

Clearly $M$ has rank $\geq 1$. Equivalently, the rank of $M$ is the maximal $n$ such that $\mathbb{R}^n$ acts freely in $M$.

**Theorem 1.5 (The rank theorem, [Lima (1965)]).** *The 3-sphere $S^3$ and the product $S^2 \times S^1$ are rank 1 manifolds. The rank of the 3-torus $T^3$ is 3.*

We shall return to this result in Chapter 7 (see Theorem 7.1).

### 1.2.5  *Turbulization*

Let $\mathcal{F}$ a codimension one foliation on a 3-manifold and $\gamma$ be a closed curve such that $\gamma$ is transverse to $\mathcal{F}$. we assume that $\gamma$ is orientable, (*i.e.*, it has a solid torus tubular neighborhood). We modify $\mathcal{F}$ along $\gamma$ as follows. Pick a neighborhood $U$ of $\gamma$ and suppose that $U$ is diffeomorphic to a solid torus $S^1 \times D^2$. Since $\gamma$ is transverse to $\mathcal{F}$ we can assume that $\mathcal{F}$ intersects the solid torus in the trivial foliation by discs $\theta \times D^2$, $\theta \in S^1$.

We consider the Reeb foliation $\mathcal{F}_R$ in $S^1 \times D^2$. We replace (by surgery) the foliation $\mathcal{F} \cap U$ by $\mathcal{F}_R$ in $U$ to obtain a foliation $\mathcal{F}_\gamma$ as in Figure 1.12.

The resulting foliation $\mathcal{F}_\gamma$ is said to be obtained by turbulization of $\mathcal{F}$ along $\gamma$. Note that $\mathcal{F}_\gamma$ is $C^r$ if $\mathcal{F}$ is, $0 \leq r \leq \infty$, $r \neq w$. The number of Reeb components of the new foliation $\mathcal{F}_\gamma$ is greater than or equal to the number of components of $\mathcal{F}$.

**Example 1.8.** Consider the foliation by discs $\theta \times D^2$ in $S^1 \times D^2$, and let $\gamma = S^1 \times \{0\}$ be the curve in the middle of $S^1 \times D^2$. We have that $\gamma \pitchfork \mathcal{F}$,

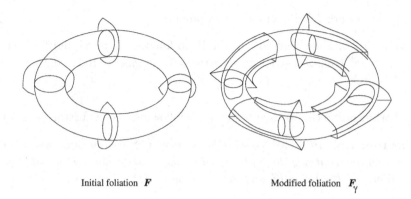

Initial foliation **F**                    Modified foliation **F**$_\gamma$

Fig. 1.12    Turbulization.

and so, we can modify $\mathcal{F}$ by turbulization along $\gamma$. The compact leaves of the resulting foliation are diffeomorphic to the torus $T^2$. The non-compact ones are all either planes $\mathbb{R}^2$ (inside of the Reeb component) or punctured discs $D^2\backslash\{\text{point}\}$ (outside of the Reeb component).

**Example 1.9.** Consider $\mathcal{F}$ as before and $\gamma$ as in Figure 1.13.

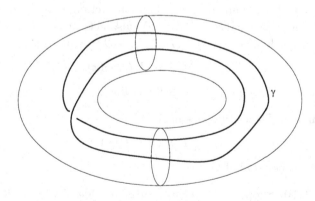

Fig. 1.13

Note that $\gamma \pitchfork \mathcal{F}$ and so we can modify $\mathcal{F}$ by turbulization along $\gamma$. In this case, $\mathcal{F}_\gamma$ is a foliation of $S^1 \times D^2$ whose leaves are $T^2$, $\mathbb{R}^2$ and $D^2 - \{2 \text{ points}\}$. Analogously, it is easy to construct a foliation of $S^1 \times D^2$

whose leaves are $T^2$, $\mathbb{R}^2$ and $D^2 - \{n \text{ points}\}$.

**Exercise 1.2.6.** Show that the for the Reeb foliation $\mathcal{F}$ in $S^3$ and for every curve $\gamma \subset S^3$ transverse to $\mathcal{F}$ we have the same number of Reeb components for $\mathcal{F}$ and $\mathcal{F}_\gamma$.

Turbulization can be used to prove the following (see [Thurston (1975)]):

**Theorem 1.6.** *All closed 3-manifolds support $C^\infty$ codimension one foliations. On the contrary, the only closed surfaces admitting codimension one foliations are the torus $T^2$ and the Klein bottle $K^2$.*

### 1.2.6 *Suspensions*

A *representation* of a group $G$ in a group $H$ is a homomorphism

$$Q \colon G \to H.$$

We shall be interested in the case $G = \pi_1(B)$ and $H = \mathrm{Diff}^r(F)$ where $B, F$ are manifolds and $\mathrm{Diff}^r(F)$ is the group of class $C^r$ diffeomorphisms in $F$ endowed with the composition operation.

Suppose that $Q \colon \pi_1(B) \to \mathrm{Diff}^r(F)$ is a representation of $\pi_1(B)$ in $\mathrm{Diff}^r(F)$. Let $\widetilde{B} \xrightarrow{\pi} B$ be the universal covering of $B$. Recall that $\pi_1(B)$ acts in $\widetilde{B}$ by deck transformations: $\alpha \in \pi_1(B)$, $\tilde{b} \in \widetilde{B}$, $b = \pi(\tilde{b})$, we have $\tilde{\alpha}$ the lift of $\alpha$. Define $\alpha \cdot \tilde{b} = \tilde{\alpha}(1)$ as the action of $\pi_1(B)$ in $\widetilde{B}$. With this action one has $\widetilde{B}/\pi_1(B) \simeq B$. $\pi_1(B)$ also acts in $\widetilde{B} \times F$ via $Q$ in the following way: define $A \colon \pi_1(B) \times (\widetilde{B} \times F) \to \widetilde{B} \times F$ by setting

$$A(\alpha, (\tilde{b}, x)) = (\alpha \cdot \tilde{b}, Q(\alpha)(x)).$$

$B \times_Q F = (\widetilde{B} \times F)/A \to B \times_Q F$ is a manifold.

**Definition 1.9.** The orbit space of $A$,

$$B \times_Q F = (\widetilde{B} \times F)/A$$

is called the *suspension* of $Q$.

**Example 1.10.** Suppose that $Q(g) = \mathrm{Id}_F$ (the identity in $F$) for all $g \in \pi_1(B)$. Then $B \times_Q F$ is precisely the cartesian product $B \times F$.

**Example 1.11.** Suppose $B = S^1$. In this case $\widetilde{B} = \mathbb{R}$ and $\pi_1(B) = \mathbb{Z}$. Let $Q \colon \pi_1(B) \to \mathrm{Diff}^r(F)$ be a representation. Then $Q(n) = (f^{-1})^n = f^{-1} \circ f^{-1} \circ \cdots \circ f^{-1}$, with $f \colon F \to F$ being a diffeomorphism of $F$. By definition $A(n, (\tilde{b}, x)) = (\tilde{b} + n, f^{-n}(x))$. Note that $A$ identifies $(0, x)$ with

$(0 + 1, f^{-1}(x)) = (1, f^{-1}(x))$ by replacing $x$ by $f(x)$, we have that $A$ identifies $(0, f(x))$ with $(1, x)$.

Note that $S^1 \times_Q F$ exhibits a flow given by projecting the constant flow $\dfrac{\partial}{\partial t}$ onto $\mathbb{R} \times F$. See Figure 1.14.

The suspension $B \times_Q F$ is equipped with two foliations $\mathcal{F}_Q$, $\mathcal{F}'_Q$ defined as follows: The action $A$ leaves invariant the horizontal and vertical foliations in $\widetilde{B} \times F$ given by $\widetilde{\mathcal{F}} = \{\widetilde{B} \times f : f \in F\}$, $\widetilde{\mathcal{F}}' = \{\widetilde{b} \times F : \widetilde{b} \in \widetilde{B}\}$.

Hence the action $A$ induces a pair of foliations $\mathcal{F}_Q, \mathcal{F}'_Q$ in $B \times_Q F$ whose leaves $L, L'$ satisfy
$$L = \pi(\text{leaf of } \widetilde{\mathcal{F}}), \quad L' = \pi(\text{leaf of } \widetilde{\mathcal{F}}'),$$
where $\pi : \widetilde{B} \times F \to B \times_Q F$ is the quotient map. Note that the foliations $\mathcal{F}_Q, \mathcal{F}'_Q$ in $\widetilde{B} \times F$ are transverse. We shall discuss more properties of these foliations later on.

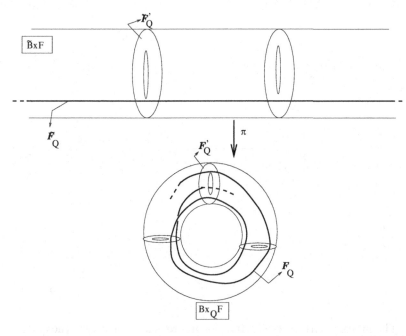

Fig. 1.14

**Example 1.12.** Let $B$ be the bitorus, *i.e.*, the genus two orientable closed surface. The fundamental group $\pi_1(B)$ has the following presentation:
$$\pi_1(B) = < a, b, c, d : aba^{-1}b^{-1}cdc^{-1}d^{-1} = 1 > .$$

Fix $f, g \in \mathrm{Diff}^r(S^1)$ and define the presentation $Q \colon \pi_1(B) \to \mathrm{Diff}^r(S^1)$ by setting

$$Q(a) = f, \quad Q(c) = g, \quad Q(b) = Q(d) = \mathrm{Id}$$

and extending linearly. The projection $Q$ is well defined since

$$Q(aba^{-1}b^{-1}cdc^{-1}d^{-1}) = 1.$$

Let us describe the suspension $B \times_Q S^1$ of $Q$. On one hand, consider the subgroup $G$ of $\pi_1(B)$ generated by $b$ and $d$, *i.e.*, $G = < b, d >$. On the other hand observe that the universal covering of $B$, $\tilde{B}$, is the Poincaré disc. Let $A_G \colon G \times \tilde{B} \times S^1 \to \tilde{B} \times S^1$ the action $A$ restricted to $G$, namely $A_G(\tilde{x}, \theta) = (g \cdot x, Q(g)(\theta))$. Clearly $Q = \mathrm{Id}$ in $G$ and so

$$(\tilde{B} \times S^1)/A_G = \tilde{B}/G \times S^1.$$

Consider $S^1$ as the unit interval $[0, 1]$ with $0 \approx 1$. Figure 1.15 describes the orbit space $(\tilde{B} \times S^1)/A_G$ of $A_G$.

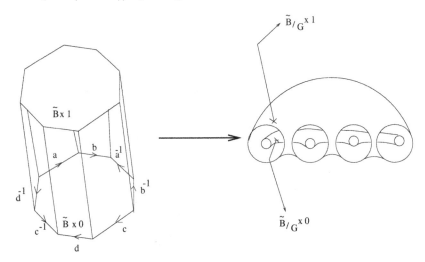

Fig. 1.15

The internal surface $\tilde{B}/G \times 0$ in the figure is identified with the external one $\tilde{B}/G \times 1$. To obtain $B \times_Q S^1$ we identify the intermediate curves $a \times g(\theta), a^{-1} \times \theta, c \times f(\theta), c^{-1} \times \theta$ according to Figure 1.16.

The leaves of the resulting foliation $\mathcal{F}_Q$ spiral around the suspended manifold according to the maps $f, g$. The other foliation $\mathcal{F}_Q'$ yields a foliation by circles of $B \times_Q S^1$, and so, $B \times_Q S^1$ is Seifert. We shall be back to this example later on.

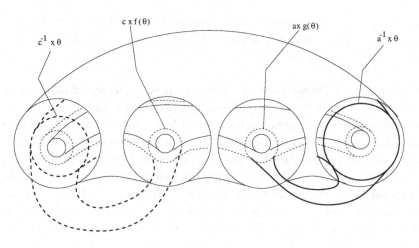

$c \times f(\theta)$

$\bar{c}^1 \times \theta$

$ax\, g(\theta)$

$\bar{a}^1 \times \theta$

Fig. 1.16

### 1.2.7 *Foliations transverse to the fibers of a fiber bundle*

In this section we discuss an important class of foliations given by suspensions, the class of foliations transverse to the fibers of a fiber bundle. Let us first recall some basic definitions:

**Example 1.13 (Fibre bundle).** A (differentiable) *fiber bundle* over a manifold $M$ is given by a differentiable map $\pi\colon E \to M$ from a manifold $E$, called *total space*, which is (the map) a submersion having the following *local triviality property*: for any $p \in M$ there exist a neighborhood $p \in U \subset M$ and a diffeomorphism $\varphi_U\colon \pi^{-1}(U) \subset E \underset{\sim}{\longrightarrow} U \times F$, where $F$ is fixed manifold called *typical fiber* of the bundle, such that the following diagram commutes

$$\begin{array}{ccc} \pi^{-1}(U) & \xrightarrow{\varphi_U} & U \times F \\ {\scriptstyle \downarrow \pi} & \swarrow {\scriptstyle \pi_1} & \\ U & & \end{array}$$

where $\pi_1\colon U \times F \to U$ is the first coordinate projection $\pi_1(x, f) = x$. In other words $\varphi_U$ is of the form $\varphi_U(\tilde{x}) = (\pi(\tilde{x}), \dots)$. Such a diffeomorphism $\varphi_U$ is called a *local trivialization* of the bundle and $U$ is a *distinguished neighborhood* of $p \in M$. Given $p \in M$ the *fiber over $p$* is $\pi^{-1}(p) \subset E$ and by the local trivialization each fiber is an embedded submanifold diffeomorphic to $F$.

According to the theorem of Ehresmann (Theorem 1.10) any $C^2$ proper submersion defines a fiber bundle as above. Let us motivate our next definition with an example.

**Example 1.14 (Suspension of a foliation by a group of diffeomorphisms).** A well known way of constructing transversely homogeneous foliations on fibered spaces, having a prescribed *holonomy group* (see Chapter 4 for the definition of holonomy) is the *suspension* of a foliation by a group of diffeomorphisms. This construction is briefly described below: Let $G$ be a group of $C^r$ diffeomorphisms of a differentiable manifold $N$. We can regard $G$ as the image of a representation $h: \pi_1(M) \to \mathrm{Diff}^r(N)$ of the fundamental group of a complex (connected) manifold $M$. Considering the differentiable universal covering of $M$, $\pi: \widetilde{M} \to M$ we have a natural free action $\pi_1: \pi_1(M) \times \widetilde{M} \to \widetilde{M}$, *i.e.*, $\pi_1(M) \subset \mathrm{Diff}^r(\widetilde{M})$ in a natural way. Using this we define an action $H: \pi_1(M) \times \widetilde{M} \times N \to \widetilde{M} \times N$ in the natural way: $H = (\pi_1, h)$. The quotient manifold $\frac{\widetilde{M} \times N}{H} = M_h$ is called the *suspension manifold* of the representation $h$. The group $G$ appears as the *global holonomy* of a natural foliation $\mathcal{F}_h$ on $M_h$ (see [Camacho and Lins-Neto (1985)]). We shall explain this construction in more details. Let $M$ and $N$ be differentiable manifolds of class $C^r$. Denote by $\mathrm{Diff}^r(N)$ the group of $C^r$ diffeomorphisms of $N$. Given a representation of he fundamental group of $M$ in $\mathrm{Diff}^r(N)$, say $h: \pi_1(M) \to \mathrm{Diff}^r(N)$, we will construct a differentiable fiber bundle $M_h$, with base $M$, fiber $N$, and projection $P: M_h \to M$, and a $C^r$ foliation $\mathcal{F}_h$ on $M_h$, such that the leaves of $\mathcal{F}$ are transverse to the fibers of $P$ and if $L$ is a leaf of $\mathcal{F}$ then $P|_L: L \to M$ is a covering map. We will use the notation $G = h(\pi_1(M)) \subset Aut(N)$.

Let $\pi: \widetilde{M} \to M$ be the $C^r$ universal covering of $M$. A covering automorphism of $\widetilde{M}$ is a diffeomorphisms $f$ of $\widetilde{M}$ that satisfies $\pi \circ f = \pi$. If we consider the natural representation $g: \pi_1(M) \to Aut(\widetilde{M})$ then we know that:

(a) $g$ is injective. In particular $g(\pi_1(M))$ is isomorphic to $\pi_1(M)$.

(b) $g$ is properly discontinuous.

We can therefore define an action $H: \pi_1(M) \times \widetilde{M} \times N \to \widetilde{M} \times N$ in a natural way:

If $\alpha \in \pi_1(M)$, $\tilde{m} \in \widetilde{M}$ e $n \in N$ then

$$H(\alpha, \tilde{m}, n) = (g(\alpha)(\tilde{m}), h(\alpha)(n)).$$

Using (b) it is not difficult to see that $H$ is properly discontinuous. Thus, the orbits of $H$ define an equivalence relation in $\widetilde{M} \times N$, whose corresponding quotient space is a differentiable manifold of class $C^r$.

**Definition 1.10.** The manifold $\frac{\widetilde{M} \times N}{H} = M_h$ is called the *suspension manifold* of the representation $h$.

Notice that $M_h$ is a $C^r$ fiber bundle with base $M$ and fiber $N$, whose projection $P\colon M_h \to M$ is defined by

$$P(O(\tilde{m}, n)) = \pi(\tilde{m})$$

where $O(\tilde{m}, n)$ denotes the orbit of $(\tilde{m}, n)$ by $H$.

Let us see how to construct the foliation $\mathcal{F}_h$. Consider the product foliation $\widetilde{\mathcal{F}}$ of $\widetilde{M} \times N$ whose leaves are of the form $\widetilde{M} \times \{n\}$, $n \in N$. It is not difficult to see that $\widetilde{\mathcal{F}}$ is $H$-invariant and therefore it induces a foliation of class $C^r$ and codimension $q = \dim(N)$, $\mathcal{F}_h$ on $M_h$, whose leaves are of the form $P(\widetilde{L})$, where $\widetilde{L}$ is a leaf of $\widetilde{\mathcal{F}}$.

**Definition 1.11.** $\mathcal{F}_h$ is called the *suspension foliation* of $\mathcal{F}$ by $h$.

The most remarkable properties of this construction are summarized in the proposition below (see [Godbillon (1991)], [Camacho and Lins-Neto (1985)]):

**Proposition 1.1.** *Let $\mathcal{F}_h$ be the suspension foliation of a representation $h\colon \pi_1(M) \to \mathrm{Diff}^r(N)$. Then:*

(i) *$\mathcal{F}_h$ is transverse to fibers of $P\colon M_h \to M$. Moreover, each fiber of $P$ cuts all the leaves of $\mathcal{F}_h$.*

(ii) *The leaves of $\mathcal{F}_h$ correspond to the orbits of $h$ in $N$ in a 1-to-1 correspondence.*

(iii) [2] *If $L$ is a leaf of $\mathcal{F}_h$ corresponding to the orbit of a point $p \in N$, then $P|_L\colon L \to M$ is a covering map (here $L$ is equipped with its natural intrinsic structure).*

  *This implies that fixed a point $p \in M$ and its fiber $N_p = P^{-1}(p)$, we obtain by lifting of paths in $\pi_1(M, p)$, to the leaves of $\mathcal{F}_h$, a group $G_p \subset \mathrm{Diff}^r(N_p)$, which is conjugate to $G$.*

(iv) *There exists a collection $\{y_i\colon U_i \to N\}_{i \in I}$ of submersions defined in open subsets $U_i$ of $M_h$ such that*

(a) $M_h = \bigcup_{i \in I} U_i$

(b) *$\mathcal{F}_h|_{U_i}$ is given by $y_i\colon U_i \to N$.*

---

[2]Due to (iii) we call $G$ the *global holonomy* of the suspension foliation $\mathcal{F}_h$.

(c) *if $U_i \cap U_j \neq \phi$ then $y_i = f_{ij} \circ y_j$ for some $f_{ij} \in G$.*
(d) *if $L$ is the leaf of $\mathcal{F}_h$ through the point $q \in N_p$, then the holonomy group of $L$ is conjugate to the subgroup of germs at $q$ of elements of the group $G = h(\pi_1(M,p))$ that fix the point $q$.*

Conditions (i) and (ii) above motivate the following definition:

**Definition 1.12.** Let $\xi: = (\pi: E \xrightarrow{F} B)$ be a fiber bundle. A foliation $\mathcal{F}$ on $E$ is said to be *transverse to the fibration* $\pi: E \xrightarrow{F} B$ if:
(1) $\mathcal{F}$ is transverse to each fiber of $\pi$
(2) $\dim \mathcal{F} + \dim F = \dim E$
(3) For each leaf $L \in \mathcal{F}$ the restriction $\pi|_L : L \to B$ is a covering map.
In this case $\mathcal{F}$ is conjugate to the *suspension* of the *global holonomy representation* $\varphi: \pi_1(B) \to \mathrm{Aut}(F)$ of $\mathcal{F}$. According to Ehresmann ([Camacho and Lins-Neto (1985)]) conditions (1) and (2) imply (3) when the fiber $F$ is compact.

Using the holonomy lifting paths given by condition (3) above we can easily prove:

**Theorem 1.7.** *Let $\mathcal{F}$ be a foliation of class $C^r$ transverse to the fibers of a fiber bundle $\xi: = (\pi: E \xrightarrow{F} B)$. Then $\mathcal{F}$ is conjugate to the suspension of a representation $\varphi: \pi_1(B) \to \mathrm{Aut}(F)$, indeed the global holonomy of $\mathcal{F}$ is naturally conjugate to the image $\varphi(\pi_1(B))$. Conversely if $\mathcal{F}$ is the suspension of a representation $\varphi: \pi_1(B) \to \mathrm{Aut}(F)$ for some base manifold $B$ and some fiber manifold $F$ then there is a fiber bundle space $\xi: = (\pi: E \xrightarrow{F} B)$ such that $\mathcal{F}$ is transverse to the fibers of $\xi$ and the global holonomy of $\mathcal{F}$ is conjugate to the image $\varphi(\pi_1(B)) < \mathrm{Aut}(F)$.*

Recall that a discrete finitely generated group is always conjugate to the fundamental group of a manifold. Thus, suspensions of group presentations and foliations transverse to fiber bundles are in natural bijection. As a natural complement to the above results we have:

**Theorem 1.8.** *Two representations $\varphi: \pi_1(B) \to \mathrm{Aut}(F)$ and $\tilde{\varphi}: \pi_1(\tilde{B}) \to \mathrm{Aut}(\tilde{F})$ are conjugate if, and only if, there is a fibered diffeomorphism $\Theta: E \to \tilde{E}$ (i.e, $\Theta$ is the lift of a diffeomorphism $\theta: B \to \tilde{B}$ such that $\tilde{\pi} \circ \Theta = \theta \circ \pi$ for the projections $\pi: E \to B$ and $\tilde{\pi}: \tilde{E} \to \tilde{B}$), with the property that $\Theta$ is a conjugacy between the suspension foliations in $E$ and $\tilde{E}$ of $\varphi$ and $\tilde{\varphi}$ respectively.*

### 1.2.8  *Transversely homogeneous foliations*

Now we introduce an important class of foliations. Let $G$ be a Lie group and denote by $\mathcal{G}$ the Lie algebra of $G$. The *Maurer-Cartan* form over $G$ is the unique 1-form $w \colon TG \to \mathcal{G}$ satisfying:

(i) $w(X) = X, \forall X \in \mathcal{G}$

(ii) $Lg^*w = w, \forall g \in G$; where $Lg \colon G \hookrightarrow G$ is the left-translation $x \in G \mapsto gx \in G$, $g \in G$ fixed.

The 1-form $w$ satisfies the *Maurer-Cartan formula* $dw + \frac{1}{2}[w,w] = 0$. In fact, given $X, Y \in \mathcal{G}$ we have

$$dw(X,Y) = X.w(Y) - Y.w(X) - w([X,Y]) = -[X,Y].$$

But

$$[w,w](X,Y) = [w(X),w(Y)] - [w(Y),w(X)] = 2[X,Y]$$

because $X$ and $Y$ belong to $\mathcal{G}$ and $w(X) = X, \forall X \in \mathcal{G}$.

Thus we have $dw(X,Y) + \frac{1}{2}[w,w](X,Y) = 0, \forall X, Y \in \mathcal{G}$ which proves the Maurer-Cartan formula.

Let now $\{X_1, \ldots, X_n\}$ be a basis of $\mathcal{G}$. We have $[X_i, X_j] = \sum_k c_{ij}^k X_k$ for some constants $c_{ij}^k \in \mathbb{C}$, skew-symmetric in $(i,j)$. The $c_{ij}^k$'s are the *structure constants* of $G$ in the basis $\{X_1, \ldots, X_n\}$.

Let now $\{w_1, \ldots, w_n\}$ be the dual basis to $\{X_1, \ldots, X_n\}$, with $w_j$ left-invariant. We have $dw_k = -\frac{1}{2}\sum_{i,j} c_{ij}^k w_i \wedge w_j$ and then it is easy to see that $w = \sum_k w_k X_k$ is the Maurer-Cartan form of $G$.

We recall the following theorem of Darboux and Lie:

**Theorem 1.9 ([Godbillon (1991)] pag. 230).** *Let $\alpha$ be a differentiable 1-form on a manifold $M$ taking values on the Lie algebra $\mathcal{G}$ of $G$. Suppose $\alpha$ satisfies the Maurer-Cartan formula $d\alpha + \frac{1}{2}[\alpha, \alpha] = 0$. Then $\alpha$ is locally the pull-back of the Maurer-Cartan form of $G$ by a differentiable map. Moreover the pull-back is globally defined if $M$ is simply-connected; and two such local maps coincide up to a left translation of $G$.*

As an immediate corollary we have:

**Corollary 1.1.** *Let $\alpha_1, \ldots, \alpha_n$ be linearly independent differentiable 1-forms on a manifold $M$. Assume that we have $d\alpha_k = -\frac{1}{2}\sum_{i,j} c_{ij}^k \alpha_i \wedge \alpha_j$ where the $c_{ij}^k$'s are the structure constants of a Lie group $G$ in the basis*

$\{X_1, \ldots, X_n\}$. *Then, locally, there exist differentiable maps* $\pi\colon U \subset M \to G$ *such that* $\alpha_j = \pi^* w_j$, $\forall\, j$ *where* $\{w_1, \ldots, w_n\}$ *is the dual (left-invariant) basis of* $\{X_1, \ldots, X_n\}$. *Moreover if* $M$ *is simply-connected then we can take* $U = M$ *and if* $\pi\colon U \to G$, $\overline{\pi}\colon \overline{U} \to G$ *are two such maps with* $U \cap \overline{U} \neq \phi$ *and connected then we have* $\overline{\pi} = Lg \circ \pi$ *for some left-translation* $Lg$ *of* $G$.

This way we may construct foliated actions of Lie groups on manifolds by defining suitable integrable systems of 1-forms on the manifold. This gives rise to the notion of transversely homogeneous foliations which is a very important notion in the theory.

**Definition 1.13 (Transversely homogeneous foliation).** A foliation $\mathcal{F}$ has a *homogeneous transverse structure* if there are a complex Lie group $G$, a connected closed subgroup $H < G$ such that $\mathcal{F}$ admits an atlas of submersions $y_j\colon U_j \subset M \to G/H$ satisfying $y_i = g_{ij} \circ y_j$ for some locally constant map $g_{ij}\colon U_i \cap U_j \to G$ for each $U_i \cap U_j \neq \emptyset$. In other words, a suitable atlas of submersions for $\mathcal{F}$ has transition maps given by left translations on $G$ and submersions taking values on the homogeneous space $G/H$. We shall say that $\mathcal{F}$ is transversely homogeneous *of model* $G/H$.

**Example 1.14.** Let $F = G/H$ be a homogeneous space of a complex Lie group $G$ ($H \lhd G$ is a closed Lie subgroup). Any homomorphism representation $\varphi\colon \pi_1(N) \to \mathrm{Diff}(F)$ gives rise to a foliation $\mathcal{F}_\varphi$ on $(\widetilde{N} \times F)/\Phi = M_\varphi$ which is transversely homogeneous of model $G/H$.

**Example 1.15.** $G = \mathrm{Aff}(\mathbb{R}^n) = \mathrm{GL}_n(\mathbb{R}) \times \mathbb{R}^n$ acts on $\mathbb{R}^n$ by $(A, B), X \mapsto AX + B$ and the isotropy subgroup of $0 \in \mathbb{R}^n$ is $\mathrm{GL}_n(\mathbb{R}) \times 0 = H$ so that $G/H \cong \mathbb{R}^n$ and then the transversely homogeneous foliations of type $\mathrm{Aff}(\mathbb{R}^n)/\mathrm{GL}_n(\mathbb{R})$ are the *transversely affine* foliations.

**Example 1.16.** The real projective unimodular group $G = \mathrm{PSL}(2, \mathbb{R})$ acts on $\mathbb{R}P(1)$ by

$$\left( \begin{pmatrix} x & u \\ y & v \end{pmatrix}, z \right) \longmapsto \frac{xz + u}{yz + v}$$

and the isotropy subgroup of $0 \in \mathbb{R}$ is naturally identified with $H = \mathrm{Aff}(\mathbb{R})$, so that $G/H \cong \mathbb{R}P(1)$ and then the transversely homogeneous foliations of type $\mathrm{PSL}(2, \mathbb{R})/\mathrm{Aff}(\mathbb{R})$ are the *transversely projective* foliations.

Now we introduce the concept of development of a transversely homogeneous foliation which is a basic tool in the study of these foliations:

**Proposition 1.2.** *Let $\mathcal{F}$ be a transversely homogeneous foliation of model $G/H$ on $M$. Then there exist a homomorphism $h\colon \pi_1(M) \to G$, a transitive covering space $p\colon P \to M$ corresponding to the kernel $H = \mathrm{Ker}(h) \subset \pi_1(M)$ and a submersion $\Phi\colon P \to G/H$ satisfying:*

- (i) $\Phi$ *is $h$-equivariant which means that $\Phi(\alpha \circ x) = h(x) \circ \Phi$, $\forall x \in P$, $\forall \alpha \in \pi_1(M)$.*
- (ii) *The foliation $p^*\mathcal{F}$ coincides with the foliation defined by the submersion $\Phi$.*

Such a construction is called a *development of the foliation $\mathcal{F}$* (see [Godbillon (1991)] page 209 for a detailed definition).

We will give an idea of the proof of the above Proposition 1.2 according to [Godbillon (1991)]:

Let $\{y_i\colon U_i \to G\}_{i \in I}$ be a homogeneous transverse structure for $\mathcal{F}$ in $M$. Denote by $f_{ij}$ the transformation $f_{ij}\colon G/H \to G/H$ such that $y_i = f_{ij} \circ y_j$ in $U_i \cap U_j \neq \phi$.

We can identify $f_{ij}$ in a natural way as an element of $G$. Now let $E$ be the space obtained as the sum of the $U_i \times G$, $i \in I$. Denote by $G_1$ the subgroup of $G$ generated by the $f_{ij}$'s. Consider in $E$ the equivalence relation identifying $(x, y) \in U_i \times G$, where $x \in U_i \cap U_j$, with $(x, f_{ij} \circ g) \in U_j \times G$.

Denote by $P$ the quotient space $E/\sim$. Then $P$ is a principal fiber bundle $p\colon P \to M$ having a discrete structural group $G_1 \subset G$, $P$ being defined by the cocycle $(U_i, f_{ij})$. The transitive covering space $p\colon P \to M$ has $G_1$ as group of automorphisms so that there is a natural homomorphism $h\colon \pi_1(M) \to G_1 \subset G$.

Now in each $U_i \times G$ we can construct a holomorphic submersion $\Phi_i\colon U_i \times G \to G/H$ by $\Phi_i(x, g) = g(y_i(x))$. The submersion $\Phi\colon P \to G/H$ is constructed by gluing the submersions $\Phi_i$. Finally we remark that if $P$ is not connected we can replace this space by one of its connected components. □

**Corollary 1.2.** *Let $\mathcal{F}$ be a non-singular transversely homogeneous foliation on a simply-connected manifold $M$. Then $\mathcal{F}$ is given by a smooth submersion $f\colon M \to G/H$.*

**Proof.** This corollary is a straightforward consequence of the Darboux-Lie theorem above but can also be proved by the use of Proposition 1.2:

In fact, if $M$ is simply connected in Proposition 1.2 then we have $H = \text{Ker}(h) \triangleleft \pi_1(M) = 0$ so that $H = 0$ and then $P = M$. Thus Corollary 1.2 follows from ii) of this same proposition.          □

**Remark 1.1.**
(i) $\alpha \in \pi_1(M)$ acts over $P$ in the following way: Given $x \in P$ we define $\alpha \cdot x$ as the end-point of the lifting $\tilde{\alpha}_x$ of the path $\alpha_x$ based at the point $p(x)$.
(ii) Conditions (i) and (ii) in the statement of Proposition 1.2 (equivariance conditions) are essential in the theory of transversely homogeneous foliations.

Exercise 1.2.7. Give a demonstration of Darboux-Lie Theorem (Theorem 1.9) according to the following suggestion:

Given 1-forms forming a basis $\{\omega_1, ..., \omega_n\}$ of the Lie Algebra of the Lie group $G$ and given 1-forms $\{\Omega_1, ..., \Omega_n\}$ a rank-$n$ system of 1-forms in a manifold $M$ such that $d\Omega_k = \sum_{i,j} c_{ij}^k \Omega_i \wedge \Omega_j$, where the $\{c_{ij}^k\}$ are the structure constants of the Lie Algebra relatively to the given basis, we can define 1-forms $\Theta_j = \Omega_j - \omega_j, j = 1, ..., n$; in a natural way in the product manifold $M \times G$. The system $\{\Theta_1, ..., \Theta_n\}$ is integrable and by Frobenius Theorem defines a foliation $\mathcal{F}$ of the product manifold. Given a leaf $L \in \mathcal{F}$ we have that $\Omega_j$ and $\omega_j$ coincide over $L$. Using then the natural projections $M \times G \to M$ and $M \times G \to G$ we can obtain local submersions $\pi \colon U \subset M \to G$ such that $\pi^* \omega_j = \Omega_j, \forall j$. In order to conclude one has to prove that if a diffeomorphism $\xi$ of $G$ preserves $\omega_j$ for all $j$ then $\xi$ is a left translation in $G$.

### 1.2.9    *Fibrations and the theorem of Ehresmann*

The fibers of the bundle are the leaves of a foliation on $E$. Such a foliation is also called a *fibration*. This situation is quite usual as shown in the following result:

**Theorem 1.10 (Ehresmann).** *Let $f \colon M \to N$ be a $C^2$ submersion which is a proper map, (i.e., $f^{-1}(K) \subset M$ is compact $\forall K \subset N$ compact). Then $f$ defines a fiber bundle over $N$.*

**Proof.** The proof is based in the construction of suitable compactly supported vector fields. Let $q \in M$ be given and let $F := \pi^{-1}(q) \subset E$. Then $F$ is a compact submanifold of $E$. Choose local coordinates $(t_1, \ldots, t_m)$ in a neighborhood $U$ of $q$ in $M$, with $t_j(q) = 0$, $j = 1, \ldots, m$. We take $U$ small enough so that we have:

(i) $\pi^{-1}(U)$ is relatively compact (recall that $\pi$ is proper) in $E$.

(ii) There exist smooth vector fields $X_1, \ldots, X_m$ in $\pi^{-1}(U)$ such that
$$\pi_*(X_j) = \frac{\partial}{\partial t_j}$$

**Claim 1.2.** *We have* $\pi(\Psi(y,p)) = y \quad \forall (y,p) \in V_1 \times F$.

**Proof.** Given $y \in V_1$ and $p \in F$ denote by $\gamma(z)$ the solution of the ordinary differential equation $\gamma'(z) = Z_y(\gamma(z))$ with initial condition $\gamma(0) = p$ which is defined for all $z \in \mathbb{D}(2)$. Then $\Psi(y,p) = \gamma(1)$ by definition. We have

$$\gamma'(z) = Z_y(\gamma(z)) = \sum_{j=1}^{m} t_j(y) \cdot X_j(\gamma(z)).$$

Therefore

$$\pi_*(\gamma'(z)) = \sum_{j=1}^{m} t_j(y) \cdot \frac{\partial}{\partial t_j}, \quad \text{that is,}$$

$$\frac{d}{dz}((\pi \circ \gamma)(z)) = \sum_{j=1}^{m} t_j(y) \cdot \frac{\partial}{\partial t_j} \quad \text{in} \quad \mathbb{R}^m.$$

Therefore, $(\pi \circ \gamma)(z) = (\pi \circ \gamma)(0) = z \cdot (t_1(y), \ldots, t_m(y))$ and then
$(\pi \circ \gamma)(1) = \pi(p) + (t_1(y), \ldots, t_m(y)) \Rightarrow$ (since $\pi(p)$ corresponds to the origin and $(t_1(y), \ldots, t_m(y))$ to $y$ in the local chart $(t_1, \ldots, t_m)$)
$\pi(\gamma(1)) = y \quad$ and therefore quad $\pi(\Psi(y,p)) = y$. $\qquad\square$

It remains to prove that $\Psi(V_1 \times F) = \pi^{-1}(V_1)$ for sufficiently small $V_1 \ni q$. Since $\pi \circ \Psi = \pi_2$ we have $\pi(\Psi(V_1 \times F)) \subseteq V_1$ so that $\Psi(V_1 \times F) \subseteq \pi^{-1}(V_1)$. If we do not have equality for sufficiently small $V_1$ then we obtain a sequence $q_n \in U$ with $q_n \to q$ and such that $\pi^{-1}(q_n)$ contains some point $p_n$ which does not belong to the image of $\Psi$ and in fact $\{p_n\}$ avoids some neighborhood $W$ of $F$ in $E$. Therefore, since $\Psi$ is proper, $\{p_n\}$ has some convergent subsequence say, $p_{n_j} \xrightarrow[j \to \infty]{} p$. But this implies $\pi(p_{n_j}) \xrightarrow[j \to \infty]{} \pi(p)$.

For any point $y \in U$ we consider the vector field $Z_y := t_1(y).X_1 + \cdots + t_m(y).X_m$, defined in $\pi^{-1}(U)$. In particular $Z_q = 0$ and its flow is complete (defined for all real time). Since $Z_y$ depends differentiably on $y \in U$ we have the following:

**Lemma 1.1.** *There exists a neighborhood* $q \in V \subset U$ *such that:*

(i) *for each $y \in V$, the flow of $Z_y$ is defined in*

$$\mathbb{D}(2) \times \pi^{-1}(V) \quad (\text{where } \mathbb{D}(2) = \{z \in \mathbb{C} \, ; |z| < 2\}),$$

*giving a smooth map*

$$\begin{aligned} \varphi^y : \mathbb{D}(2) \times \pi^{-1}(V) &\to \pi^{-1}(U) \\ (t, p) &\mapsto \varphi^y(t, p) \end{aligned} \quad (\text{where } t \text{ is the real time})$$

*with $\varphi^y(0, p) = p, \quad \forall p \in \pi^{-1}(V)$,*

$$\left. \frac{\partial}{\partial t} \varphi^y(t, p) \right|_{(t=0)} = Z_y(\varphi^y(t_0, p)).$$

(ii) *For some neighborhood $q \in V_1 \subset V$ we have $\varphi^y(t, p) \in V, \quad \forall p \in V_1$, $\forall t$ with $|t| \leq 1$.*

Now we may consider the time one flow map

$$\Psi : V_1 \times F \to E, \quad \Psi(y, p) := \varphi^y(1, p) \in U.$$

Then $\psi$ is holomorphic and we have an inverse for $\psi$, which is given by

$$\Psi^{-1} : \Psi(V_1 \times F) \to V_1 \times F, \quad \Psi^{-1}(p) := \varphi^y(-1, p).$$

This inverse is well-defined because of (i) and (ii) above so that $q_{n_j} \to \pi(p)$ and $\pi(p) = q$. Thus $p \in F$ what is not possible for $p_n \in E \backslash W, \quad \forall n$. This contradiction show that we must have $\Psi(V_1 \times F) = \pi^{-1}(V_1)$ for every sufficiently small neighborhood $V_1$ of $q$ in $M$. $\qquad \square$

This is the case for instance if $M$ is compact. One very important result concerned with this framework is due to Tischler (see Chapter 8).

**Theorem 1.11 (Tischler, [Tischler (1970)]).** *A compact (connected) manifold $M$ fibers over the circle $S^1$ if, and only if, $M$ supports a closed non-singular 1-form.*

This is the case if $M$ admits a codimension one foliation $\mathcal{F}$ which is invariant by the flow of some non-singular transverse vector field $X$ on $M$ as we will see in Chapter 8.

## 1.3 Holomorphic Foliations

A (real) manifold $M^{2n}$ is a *complex manifold* if it admits a differentiable atlas $\{\varphi_j \colon U_j \subset M \to \mathbb{R}^{2n}\}_{j \in J}$ whose corresponding change of coordinates are holomorphic maps $\varphi_j \circ \varphi_i^{-1} \colon \varphi_1(U_i \cap U_j) \subset \mathbb{R}^{2n} \simeq \mathbb{C}^n \to \varphi(U_i \cap U_j) \supset \mathbb{R}^{2n} \simeq \mathbb{C}^n$. Such an atlas is called *holomorphic*.

In this case all the basic concepts of differentiable manifolds (as tangent space, tangent bundle, etc...) can be introduced in this complex setting. This is the case of the concept of foliation:

**Definition 1.14.** A *holomorphic foliation* $\mathcal{F}$ of (complex) dimension $k$ an a complex manifold $M$ is given by a *holomorphic atlas* $\{\varphi_j \colon U_j \subset M \to V_j \subset \mathbb{C}^n\}_{j \in J}$ with the *compatibility property*: Given any intersection $U_i \cap U_j \neq \emptyset$ the change of coordinates $\varphi_j \circ \varphi_i^{-1}$ preserves the horizontal fibration on $\mathbb{C}^n \simeq \mathbb{C}^k \times \mathbb{C}^{n-k}$.

Examples of such foliations are, like in the "real" case, given by non-singular holomorphic vector fields, holomorphic submersions, holomorphic fibrations and locally free holomorphic complex Lie group actions on complex manifolds.

**Remark 1.2.** (i) As in the "real" case, the study of holomorphic foliations may be very useful in the classification Theory of complex manifolds.

(ii) In a certain sense, the "holomorphic case" is closer to the "algebraic case" than the case of real foliations.

### 1.3.1 *Holomorphic foliations with singularities*

One of the most common compactifications of the complex affine space $\mathbb{C}^n$ is the complex projective space $\mathbb{C}P(n)$. It is well-known that any foliation (holomorphic) of codimension $k \geq 1$ on $\mathbb{C}P(n)$ must have some *singularity* (in other words, $\mathbb{C}P(n)$, for $n \geq 2$, exhibits no holomorphic foliation in the sense we have considered up to now.) Thus one may consider such objects: *singular holomorphic foliations* as part of the zoology. Thus one may have consider "singular foliations" when dealing with complex settings.

**Example 1.17 (Polynomial vector fields on $\mathbb{C}^2$).** Let $X$ be a polynomial vector field on $\mathbb{C}^2$, say $X = P(x,y)(\partial/\partial x) + Q(x,y)(\partial/\partial y) = (P, Q)$. We have an ordinary differential equations:

$$\begin{cases} \dot{x} = P(x,y) \\ \dot{y} = Q(x,y) \end{cases}$$

We have local solutions given by the theorem of existence and uniqueness of Picard for ordinary differential equations ([Hirsch and Smale (1974)]):

$$\varphi(z) = (x(z), y(z))$$

$$\frac{d\varphi}{dz} = \dot{\varphi}(z) = X(\varphi(z))$$

Gluing the images of these unique local solutions, we can introduce the *orbits* of $X$ on $\mathbb{C}^2$. The orbits are immersed Riemann surfaces on $\mathbb{C}^2$, which are locally given by the solutions of $X$.

Now we may be interested in what occurs these orbits in "a neighborhood of the infinity $L_\infty$". We may for instance compactify $\mathbb{C}^2$ as the projective plane $\mathbb{C}P(2) = \mathbb{C}^2 \cup L_\infty$, $L_\infty \cong \mathbb{C}P(1)$. Some natural questions are then:

- What happens to $X$ in a neighborhood of $L_\infty$?
- Is it still possible to study the orbits of $X$ in a neighborhood of $L_\infty$?

We may rewrite $X$ as the coordinate system $(u, v)$ : $X(u, v) = \frac{1}{u^m} Y(u, v)$, $m \in \mathbb{N} \cup \{0\}$ where $Y$ is a polynomial vector field. The exterior product of $X$ and $Y$ is zero in common domain $U : X \wedge Y = 0$. So, orbits of $Y$ (or $X$) are orbits of $X$ (or $Y$), respectively in $U$. Then the orbits of $X$ *extend* to the $(u, v)$-plane as the corresponding orbits of $Y$ along $L_\infty$. This same way, we may consider in the $(r, s)$ coordinate system. These extensions are called *leaves* of a foliation induced by $X$ on $\mathbb{C}P(2)$. We obtain this way: A decomposition of $\mathbb{C}P(2)$ into immersed complex curves which are locally arrayed, as the orbits (solutions) of a complex vector field. This is a holomorphic foliation $\mathcal{F}$ with singularities of dimension one on $\mathbb{C}P(2)$.

**Definition 1.15.** Let $M$ be a complex manifold. A *singular holomorphic foliation* of codimension one $\mathcal{F}$ on $M$ is given by an open cover $M = \bigcup_{j \in J} U_j$ and holomorphic integrable 1-forms $\omega_j \in \bigwedge^1(U_j)$ such that if $U_j \cap U_j \neq \emptyset$, then $\omega_i = g_{ij}\omega_j$ in $U_i \cap U_j$, for some $g_{ij} \in \mathcal{O}^*(U_i \cap U_j)$. We put $\text{sing}(\mathcal{F}) \cap U_j = \{p \in U_j; \omega_j(p) = 0\}$ to obtain $\text{sing}(\mathcal{F}) \subset M$, a well-defined analytic subset of $M$, called singular set of $\mathcal{F}$. $M \setminus \text{sing}(\mathcal{F})$ is foliated by a holomorphic codimension one (regular) foliation $\mathcal{F}^1$.

**Remark 1.3.** We may always assume that $\text{sing}(\mathcal{F}) \subset M$ has codimension $\geq 2$. If $(f_j = 0)$ is an equation of codimension one component of $\text{sing}(\mathcal{F}) \cap U_j$, then we get $\omega_j = f_j^n \bar{\omega}_j$ where $\bar{\omega}_j$ is a holomorphic 1-form and $\text{sing}(\bar{\omega}_j)$ does not contain $(f_j = 0)$.

Using this we may also reformulate the definition above as follows:

**Definition 1.16.** A *singular holomorphic foliation* $\mathcal{F}$ of codimension one on $M$ is given by a pair $\mathcal{F} = (\mathcal{F}', \text{sing}(\mathcal{F}))$ where:

(1) $\text{sing}(\mathcal{F}) \subset M$ is an analytic subset of codimension $\geq 2$.
(2) $\mathcal{F}'$ is a regular holomorphic foliation of codimension one on $M' = M \setminus \text{sing}(\mathcal{F})$.

**Remark 1.4.** Assume that we have a holomorphic regular foliation $\mathcal{F}^1$ on $U - \setminus 0$, $0 \in \mathbb{C}^2$, $U \cap \text{sing}(\mathcal{F}) = \setminus 0$. Choose local coordinates $(x, y)$ centered at 0 and define a meromorphic function $f: U - \setminus 0 \to \overline{\mathbb{C}}$, $p \in U - \setminus 0$, as $f(p) = $ inclination of the tangent to the leaf $L_p$ of $\mathcal{F}^1$. By Hartogs' Extension Theorem [Siu (1974)],[Gunning I (1990)] $f$ extends to a meromorphic function $f: U \to \overline{\mathbb{C}}$. We may write $f(x, y) = \frac{a(x,y)}{b(x,y)}$, $a, b \in \mathcal{O}(U)$ and define

$$\frac{dy}{dx} = f(x, y) = \frac{b(x, y)}{a(x, y)},$$

that is,

$$\left\{ \begin{array}{l} \dot{x} = a(x, y) \\ \dot{y} = b(x, y) \end{array} \right\}.$$

Therefore, $\mathcal{F}$ is defined by a holomorphic 1-form $\omega = a(x, y)\, dy - b(x, y)\, dx$ in $U$.

**Example 1.18.** Let $f: M \to \overline{\mathbb{C}}$ be a meromorphic function on the complex manifold $M$. Then $\omega = df$ defines a holomorphic foliation of codimension one with singularities on $M$. The leaves are the connected components of the levels $\{f = \text{const.}\}$. The singular points of the foliation are divided into two classes: (1) points where $f$ or $1/f$ is well-defined, but has a critical point. (2) indeterminacy points of $f$.

**Example 1.19.** Let $G$ be a complex Lie group and $\varphi : G \times M \to M$ a holomorphic action of $G$ on $M$. The action is foliated if all its orbits have a same fixed dimension. In this case there exists a holomorphic regular foliation $\mathcal{F}$ on $M$, whose leaves are orbits of $\varphi$. However, usually, actions are not foliated, though they may define singular holomorphic foliations. For instance, an action $\varphi$ of $G = (\mathbb{C}, +)$ on $M$, $\varphi : \mathbb{C} \times M \to M$ is a holomorphic flows. We have a holomorphic complete vector field $X = \frac{\partial \phi}{\partial t}|_{t=0}$ on $M$. The singular set of $X$ may be assumed to be of codimension $\geq 2$ and we obtain a holomorphic singular foliation of dimension one $\mathcal{F}$ on $M$ whose leaves are orbits of $X$, or equivalently, of $\varphi$.

One very general problem is the study and classification of actions of complex Lie groups $G$ on a given compact complex $M$. One possible approach is to consider the stand-point of singular holomorphic foliations theory.

**Example 1.20 (Darboux foliations).** Let $M$ be a complex manifold and let $f_j \colon M \to \overline{\mathbb{C}}$ be meromorphic functions and $\lambda_j \in \mathbb{C}^*$ complex numbers, $j = 1, \ldots, r$. The meromorphic integrable 1-form $\omega = \prod_{j=1}^{r} f_j \sum_{i=1}^{r} \lambda_i \frac{df_i}{f_i}$ defines a *Darboux foliation* $\mathcal{F} = \mathcal{F}(w)$ on $M$. The foliation $\mathcal{F}$ has $f = \prod_{j=1}^{r} f_j^{\lambda_j}$ as a *logarithmic* first integral.

**Example 1.21 (Riccati foliations).** A *Riccati Foliation* on $\overline{\mathbb{C}} \times \overline{\mathbb{C}}$ is given in some affine chart $(x, y) \in \mathbb{C} \times \mathbb{C}$ by a polynomial 1-form $\omega = p(x)dy - (y^2 c(x) - yb(x) - a(x))dx$. Such a foliation is transverse to the fibration $\overline{\mathbb{C}} \times \overline{\mathbb{C}} \to \overline{\mathbb{C}}$ , $(x, y) \mapsto x$, except for a finite number of invariant fibers given in the affine part by $p(x) = 0$. This transversality allows to define a *global holonomy* of the horizontal projective line $\Lambda_0 = \overline{(y = 0)}$ which gives us a group of Möebius transformations $G \subset \mathrm{SL}(2; \mathbb{C})$ of a non-invariant vertical fiber. If $a \equiv 0$ then $\Lambda_0$ is an invariant divisor and $G$ is the usual holonomy of the leaf $\Lambda_0 \backslash \mathrm{sing}\, \mathcal{F}$ as defined above. In this case the elements of $G$ are of the form $f(z) = \frac{\alpha z}{1 + \beta z}$. Thus the holonomy of the leaf $\Lambda_0 \backslash \mathrm{sing}\, \mathcal{F}$ is solvable. In fact, the elements of $G$ are affine maps after the change of coordinates $Z = \frac{1}{z}$ on $\overline{\mathbb{C}}$. Using this remark it is easy to see that the foliation is transversely affine outside the invariant set $S$ given by the union of $\Lambda_0$ and the invariant vertical fibers given by the zeros of $p(x)$. If $a \not\equiv 0$ then $\mathcal{F}(\omega)$ is transversely projective outside $S = \bigcup_{p(x)=0} \{x\} \times \overline{\mathbb{C}}$, which is also invariant. We may induce a foliation on $\mathbb{C}P(2)$ with similar properties.

**Exercise 1.3.1 (Implicit ordinary differential equations).** An *algebraic implicit ordinary differential equation* in $n \geq 2$ complex variables is given by expressions:

$$(**) \ f_j(x_1, ..., x_n, x'_j) = 0$$

where $f_j(x_1, ..., x_n, y) \in \mathbb{C}[x_1, ..., x_n, y]$ are polynomials and the $(x_1, ..., x_n) \in \mathbb{C}^n$ are affine coordinates. Clearly, any polynomial vector field $X$ on $\mathbb{C}^n$ defines such an equation. In general $(**)$ defines a one-dimensional singular foliation in some algebraic variety of dimension $n$. In order to see it we begin by defining $F_j(x_1, ..., x_n, y_2, ..., y_n) := f_j(x_1, ..., x_n, y_j) \in \mathbb{C}[x_1, ..., x_n, y_2, ..., y_n]$ polynomials in $n + (n-1) = 2n - 1$ variables. Put also $S_j := \{(x, y) \in \mathbb{C}^n \times \mathbb{C}^{n-1}; F_j(x, y) = o\} \simeq \{(x_1, ..., x_n, y_j) \in \mathbb{C}^n_x \times \mathbb{C}_{y_j}; f_j(x_1, ..., x_n, y_j) = 0\} \times \mathbb{C}^{n-2} =: \Lambda_j \times \mathbb{C}^{n-2}_{(y_2, ..., \hat{y}_j, ..., y_n)}$.

We consider the projectivizations $\overline{S_j} \subset \mathbb{C}P(2n-1)$ and the complete intersection subvariety $S := \overline{S_2} \cap ... \cap \overline{S_n} \subset \mathbb{C}P(2n-1)$. Given by the differential forms $\omega_j := y_j dx_1 - dx_j$ $(j = 2, ..., n)$ on $\mathbb{C}^n \times \mathbb{C}^{n-1}$. Prove that $\{\omega_j = 0, j = 2, ..., n\}$ defines an integrable system on $S$. We say that the implicit differential equation $(*)$ is *normal* if $S$ admits a normalization (desingularization) by blow-ups $\sigma \colon \hat{S} \to S$. In particular we obtain in general a singular foliation $\mathcal{F}(**)$ of dimension one on the algebraic $n$-dimensional subvariety $S \subset \mathbb{C}P(2n-1)$. Denote by $f_1 \colon S \cap \mathbb{C}^n \to \mathbb{C}^1$ the projection in the first coordinate $f_1(x_1, ..., x_n, y_2, ..., y_n) = x_1$, and extend it to a holomorphic proper mapping $f_1 \colon \overline{S} \to \mathbb{C}P(1)$. Assume now that $S$ admits a normalization $\sigma \colon \hat{S} \to S$. Show that the foliation $\mathcal{F}(**)$ lifts to a foliation by curves $\hat{\mathcal{F}}(**)$ on $\hat{S}$ and $\hat{f}_1 = f_1 \circ \sigma$ defines a holomorphic proper mapping from $\overline{\hat{S}}$ over $\mathbb{C}P(1)$. Finally, using Stein Factorization Theorem find a splitting $\hat{f}_1 \colon \hat{S} \xrightarrow{\hat{f}} B \xrightarrow{\alpha} \mathbb{C}P(1)$ where $\alpha \colon B \to \mathbb{C}P(1)$ is a finite ramified covering and $\hat{f} \colon \hat{S} \to B$ is an extended holomorphic fibration over the compact Riemann surface $B$ such that the following diagram therefore commutes

$$
\begin{array}{ccc}
\hat{S} & \xrightarrow{\sigma} & \overline{S} \\
\hat{f} \downarrow & & \downarrow f_1 \\
B & \xrightarrow{\alpha} & \mathbb{C}P(1)
\end{array}
$$

for a map $\hat{f}_1 \colon \hat{S} \to \mathbb{C}P(1)$.

Exercise 1.3.2. Let $X_{\lambda, \mu} = \lambda x \frac{\partial}{\partial x} + \mu y \frac{\partial}{\partial y}$ be a complex vector field defined in a neighborhood of the origin $0 \in \mathbb{C}^2$. Show that $X_{\lambda, \mu}$ is transverse to the 3-spheres $S^3(0, R)$ for $R > 0$ small enough, if and only if, $\lambda/\mu \in \mathbb{C} \setminus \mathbb{R}_-$. Let now $X$ be a polynomial vector field in $\mathbb{C}^2$ and assume that the singularities of the corresponding foliation $\mathcal{F}$ on $\mathbb{C}P(2)$ are of local form $X_{\lambda,\mu}$ with $\lambda/\mu \notin \mathbb{R}$. Choose small balls $\mathbb{B}(p_j)$ around the singularities $p_j \in \mathrm{sing}(\mathcal{F})$ in $\mathbb{C}P(2)$. Show that there is a foliation $\mathcal{F}_d$ in a manifold $M_d$ with the following properties: This is a $C^\infty$ regular codimension-two real foliation $\mathcal{F}_d$ on a compact real 4-manifold $M_d$, which contains two copies of the foliated pair $(\mathbb{C}P(2) \setminus \bigcup\limits_{j=1}^{r} \overline{\mathbb{B}}(p_j), \mathcal{F}|_{\mathbb{C}P(2)\setminus \bigcup\limits_{j=1}^{r} \overline{\mathbb{B}}(p_j)})$. By Schwarz Reflection Principle the leaves of $\mathcal{F}_d$ have also natural structures of Riemann surfaces. Any Riemannian metric $g$ in $\mathbb{C}P(2)$ induces a $C^\infty$ Riemannian metric $g_d$ in $M_d$, that can be chosen to be hermitian along the leaves of $\mathcal{F}_d$. Show that the leaves of the non-singular foliation $\mathcal{F}|_{\mathbb{C}P(2)\setminus \mathrm{sing}(\mathcal{F})}$ have the same growth type than the corresponding leaves of $\mathcal{F}_d$.

Chapter 2

# Plane fields and foliations

## 2.1 Definition, examples and integrability

A *k-plane field* on a manifold $M^m$, $1 \le k \le m$, is a map $x \in M \mapsto P(x)$, such that $P(x)$ is a $k$-dimensional subspace of $T_x M$. When $k = 1$, $P$ is called *line field*. A $k$-plane field $P$ is *of class $C^r$* if each point $x \in M$ has a neighborhood $U \ni x$ where there are defined $k$ linearly independent vector fields $X^1, \ldots, X^k \colon U \xrightarrow{C^r} TU$ generating $P$ in $U$, namely $P(x) = \mathrm{Span}(X^1(x), \ldots, X^k(x))$. In this case we say that $X^1, \ldots, X^k$ *generate* $P$ in $U$.

**Example 2.1.** A $C^r$ foliation $\mathcal{F}$ of dimension $k$ defines the plane field $T\mathcal{F}(x)$ of class $C^{r-1}$ given by

$$T\mathcal{F}(x) = T_x \mathcal{F}_x.$$

The plane field $N\mathcal{F}$ given by

$$N\mathcal{F}(x) = T_x M / T_x \mathcal{F}_x$$

is called the normal plane field of $\mathcal{F}$.

**Question 2.1.** *Is any plane field $P$ of the form $P = T\mathcal{F}$ for some foliation $\mathcal{F}$? Locally the answer is yes but in general the answer is no. This question suggests the following definition.*

**Definition 2.1.** A $k$-plane field $P$ of class $C^r$ is *integrable* if $P = T\mathcal{F}$ for some $C^{r+1}$ foliation $\mathcal{F}$.

### 2.1.1 *Frobenius Theorem*

Let $X, Y$ be two vector fields in a manifold $M$ and $p \in M$ be fixed. Denote by $X_t$ the flow of $X$ and similarly $Y_t$. $X, Y \in C^r$, $r \ge 2$. We define

$X_t^*(Y)(p) = DX_{-t}(X_t(p)) \cdot Y(X_t(p))$. Note that $X_t^*(X)(p) = X(p), \quad \forall t$.

**Definition 2.2.** The *Lie bracket* of $X, Y$ is the vector field $[X, Y]$ defined by

$$L_X(Y)(p) = [X, Y](p) = \frac{d}{dt}\Big|_{t=0}(X_t^*(Y)(p)) \quad X, Y \in C^r, \ r \geq 2.$$

In local coordinates, $[X, Y]$ has the following form: Writing

$$X = \sum_i a_i \frac{\partial}{\partial x_i}, \quad Y = \sum_i b_i \frac{\partial}{\partial x_i}$$

one has

$$[X, Y] = \sum_{i,j} \left(a_i \frac{\partial b_j}{\partial x_i} - b_i \frac{\partial a_j}{\partial x_i}\right) \frac{\partial}{\partial x_j}.$$

When $X$ and $Y$ are defined in an open set of $\mathbb{R}^m$, the formula above yields

$$[X, Y] = DY(p) \cdot X(p) - DX(p) \cdot Y(p).$$

A vector field $X$ is tangent to a plane field $P$ (denoted by $X \in P$) if $X(x) \in P(x)$ for all $x \in M$.

**Definition 2.3.** A plane field $P$ is *involutive* if $X, Y \in P \Rightarrow [X, Y] \in P$.

**Lemma 2.1.** *If $\mathcal{F}$ is a foliation, then its associated plane field $T\mathcal{F}$ is involutive.*

**Proof.** Let $X$, $Y$ be two vector fields tangent to $T\mathcal{F}$. By using local coordinates defining $\mathcal{F}$ one can assume that $X, Y$ are defined in an open set of $\mathbb{R}^m$ and if $\dim \mathcal{F} = k$, then

$$X(x, y) = (f(x, y), 0), \quad Y(x, y) = (f(x, y), 0)$$

$$[X, Y] = \begin{pmatrix} \partial_x g & \partial_y g \\ 0 & 0 \end{pmatrix} \begin{pmatrix} f \\ 0 \end{pmatrix} - \begin{pmatrix} \partial_x f & \partial_y f \\ 0 & 0 \end{pmatrix} \begin{pmatrix} g \\ 0 \end{pmatrix}$$

$$= (f \cdot \partial_x f - f \cdot \partial_x f, 0).$$

Hence $[X, Y] \in T\mathcal{F}$ and the proof follows.  $\square$

**Theorem 2.1 (Frobenius' Theorem).** *Involutive plane fields are integrable.*

The converse holds by the previous lemma. Hence the following assertions are equivalent:

(1) $P$ is integrable.
(2) $P = T\mathcal{F}$.
(3) $P$ is *involutive*, (*i.e.*, $X, Y \in P \Rightarrow [X, Y] \in P$).

Since all line fields are involutive one has

**Corollary 2.1.**
*All line fields are integrable.*

**Example 2.2.** Define $M = \mathbb{R}^3$ and let $P$ the map given by $P(x, y, z) =$ Span$(X, Y)$, where $X, Y$ are the vector fields defined by $X(x, y, z) = (1 + y, y, z)$ and $Y(x, y, z) = (-y_1, 1 + y, 0)$. As $X$ and $Y$ are orthogonals and non-zero everywhere one has that $P$ is a plane field of class $C^\omega$. Let us use the Frobenius theorem to show that $P$ is not integrable. Easy computations yield

$$DY = \begin{pmatrix} 0 & -1 & 0 \\ 0 & 1 & 0 \\ 0 & 0 & 0 \end{pmatrix}, \quad DX = \begin{pmatrix} 0 & 1 & 0 \\ 0 & 1 & 0 \\ 0 & 0 & 1 \end{pmatrix}.$$

So,

$$DY \cdot X = (-y, y, 0), \quad DX \cdot Y = (1_y, 1 + y, 0).$$

Hence

$$[X, Y](x, y, z) = DY(x, y, z) \cdot X(x, y, z) - DX(x, y, z) \cdot Y(x, y, z)$$

$$= (-1 - 2y, -1, 0).$$

So $[X, Y] \in P \Leftrightarrow [X, Y] = \alpha X + \beta Y$, for some $\alpha, \beta \in \mathbb{R}$. But

$$[X, Y] = \alpha X + \beta Y \Leftrightarrow (-1 - 1y, -1, 0)$$
$$= \alpha(1 + y, y, z) + \beta(-y, 1 + y, 0)$$
$$\Leftrightarrow \begin{cases} -1 - 2y = \alpha(1 + y) - \beta y \\ -1 = \alpha y + \beta(1 + y) \\ 0 = \alpha z \end{cases}$$

Replacing by $(x, y, z) = (1, 0, 1)$ one has $\alpha = 0$, $\beta = -1$, $0 = 1$, a contradiction. We conclude that $[X, Y](1, 0, 1) \notin P(1, 0, 1)$ and then $P$ is not integrable by Frobenius's.

Regarding $(m-1)$-plane fields and foliations we mention the following result:

**Theorem 2.2 (Thurston's homotopy theorem, [Thurston (1976)]).** *Every $(m-1)$-plane field in a $m$-manifold $M^m$ is homotopic to an integrable plane field $T\mathcal{F}$, where $\mathcal{F}$ is a $C^\infty$ codimension one foliation of $M$.*

**Example 2.3 (integrable systems of differential forms).** Let $\omega_1, ..., \omega_r$ be differential 1-forms of class $C^r$ on a manifold $M$ and assume that they are linearly independent at each point $p \in M^n$. We may consider the distribution $\Delta$ of $(n-r)$-dimensional planes defined by $\Delta(p) \subset T_p M$ is

$$\Delta(p) = \{v \in T_p M, \omega_j(p) \cdot v = 0, j = 1, ..., r\}.$$

This distribution is called *integrable* if it is tangent to a $-r$ dimensional foliation $\mathcal{F}$ on $M$. According to Frobenius Integrability theorem (see also [Camacho and Lins-Neto (1985)]) this occurs if and only if the system of 1-forms is *integrable* what means that we have $d\omega_j \wedge \omega_1 \wedge ... \wedge \omega_r = 0$ for all $j = 1, ..., r$. This occurs for instance if we have a closed 1-form $\omega$ with $\omega(p) \neq 0, \forall p \in M$. In this case we have a codimension one foliation $\mathcal{F}$ on $M$ which is defined by the Pfaffian equation $\omega = 0$. The leaves of $\mathcal{F}$ are locally given by $f = cte$, where $f$ is a local primitive for $\omega$.

## 2.2 Orientability

Recall that a $k$-form $w$ in $M$ is a map

$$w \colon M \to \Lambda^k(TM)$$
$$p \mapsto w(p) \colon T_p M \times \cdots \times T_p M \to \mathbb{R}.$$

**Remark 2.1 (Criterium for orientability of manifolds).** A manifold $M^m$ is orientable if, and only if, it admits a volume form $w$, (*i.e.*, a $m$-form $w$ such that $w(p) \neq 0, \forall p \in M$).

**Definition 2.4.** A $k$-plane field $P$ in $M^m$, $1 \leq k \leq m$ is *orientable* if there is a covering $\{U_i\}$ of $M$ and $k$ continuous linearly independent vector fields $X^{1,i}, \ldots, X^{k,i} \colon U_i \to TU_i$ so that

1) $P(x) = \operatorname{span}(X^{1,i}(x), \ldots, X^{k,i}(x)), \quad \forall x \in U_i$

2) $\operatorname{Det} \begin{pmatrix} X^{1,i}(x) \ldots X^{k,i}(x) \\ X^{1,j}(x) \ldots X^{k,j}(x) \end{pmatrix} > 0, \ \forall x \in U_i \cap U_j.$

We say that $P$ is *transversely orientable* if there is an orientable plane field $P'$ on $M$ such that $TM = P \oplus P'$.

**Proposition 2.1.** *A line field $P$ in $M$ is orientable if, and only if, $P(x) = \text{Span}(X(x))$ for some continuous everywhere non-vanishing vector field $X$ on $M$.*

**Proof.** The only if part is obvious by taking the trivial covering $\{U_i\} = \{M\}$ of $M$ and $X^{1,i} = X$. Now, if $P$ orientable then there exists $\{U_i\}_{i \in I}$ open covering of $M$ and $k$ vector fields $X^i = X^{1,i} : U_i \to TU_i$ such that

1) $P(x) = \text{Span}(X^i(x)), \quad \forall\, x \in U_i$.

2) $\det \begin{pmatrix} X^i(x) \\ X^j(x) \end{pmatrix} > 0, \quad \forall\, x \in U_i \cap U_j$, i.e., $X^i(x) = a_{ij}(x), \quad X^i(x)$,

$\forall\, x \in U_i \cap U_j, \quad a_{ij}(x) > 0$.

Define $X(x) = X^i(x)/\|X^i(x)\|, \quad \forall\, x \in U_i$. Then $X$ is well defined since

$$\frac{X^i(x)}{\|X^i(x)\|} = \frac{X^j(x)}{\|X^j(x)\|}, \quad \forall\, x \in U_i \cap U_j \Leftrightarrow$$

$$\frac{X^i(x)}{\|X^i(x)\|} = \frac{a_{ij}(x)X^j(x)}{\|a_{ij}(x)\|\,\|X^j(x)\|} = \frac{X^j(x)}{\|X^j(x)\|}$$

as $a_{i,j}(x) > 0$ for all $x$. Since $X^i$ generates $P$ in $U_i$ the result follows. $\quad\square$

**Example 2.4.** Choose $M = \mathbb{R}^2 - \{0\}$. None of the line fields induced by the foliations $\mathcal{F}_1, \mathcal{F}_2$ in $M$ at Figure 2.1 is orientable. One can see this by observing how the tangent vector varies along the curve indicated at $\mathcal{F}_1$.

**Proposition 2.2.** *A manifold $M$ is orientable if, and only if, the plane field $P(x) = T_x M$ is orientable.*

**Proof.** Since $P$ is orientable there are a cover $\{U_i\}$ and vector fields $X^{1,i}, \ldots, X^{m,i} : U_i \to TU_i$ such that

$$T_x M = \text{Span}(X^{1,i}(x), \ldots, X^{m,i}(x))\, \forall\, x \in U_i$$

and $\det(X^{n,k}(x))_{1 \le n \le m}\, k = i, j > 0, \forall\, x \in U_i \cap U_j$. For each $U_i$ one choose an $m$-form $W_{U_i}$ such that if $\forall\, x \in U_i$ then $\{v_1, \ldots, v_m\}$ is a base of $T_x M$ with

$$\det \begin{pmatrix} v_1, \ldots, v_m \\ X^{1,i}(x), \ldots X^{m,i}(x) \end{pmatrix} > 0 \Leftrightarrow W_{U_i}(x)(v_1, \ldots, v_m) > 0.$$

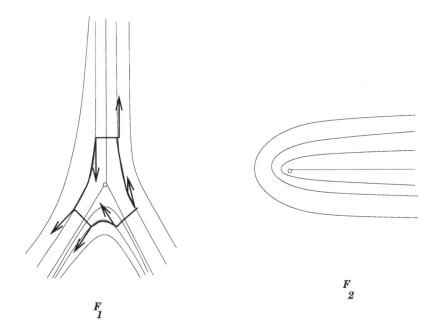

$F_1$

$F_2$

Fig. 2.1

Let $\{\phi_i\}$ be a partition of the unity subordinate to the cover $\{U_i\}$. Define $W = \sum_i \phi_i, W_{U_i}$. Then $w$ is a $m$-form with $w(x) \neq 0$, $\forall x$. In fact, for $x \subset M$, $w(x) = \sum_{\{i;x\in U_i\}} \phi_i(x)W_{U_i}(x)$. Let $i$ be such that $x \in U_i$ and $\{v_1, \ldots, v_m\}$ is a base $T_x M$ satisfying

$$\det \begin{pmatrix} v_1, \ldots, v_m \\ X^{1,i}, \ldots, X^{m,1}(x) \end{pmatrix} > 0.$$

Let $j$ be such that $x \in U_j$. Note that

$$\begin{pmatrix} v_1, \ldots, v_m \\ X^{1,j}(x), \ldots, X^{m,j}(x) \end{pmatrix} = \begin{pmatrix} X^{1,i}(x) \ldots X^{m,i}(x) \\ X^{1,j}(x) \ldots X^{m,j}(x) \end{pmatrix} \cdot \begin{pmatrix} v_1, \ldots v_m \\ X^{1,i}(x) \ldots X^{m,i}(x) \end{pmatrix}.$$

Because $P$ is orientable one has $\det \begin{pmatrix} v_1 \ldots v_m \\ X^{1,i} \ldots X^{m,1}(x) \end{pmatrix} > 0$. So $W_{U_i}(x)(v_1, \ldots, v_m) > 0$. Henceforth $\phi_i(x) \cdot W_{U_i}(x)(v_1, \ldots, v_m) \geq 0$, $\forall i$.

Then

$$W(x)(v_1, \ldots, v_m) = \sum_{\{x; x \in U_i\}} \phi_i(x) \cdot W_{U_i}(x)(v_1, \ldots, v_m) > 0$$

and so there is $i$ such that $\phi_i(x) = 1$. It follows that, $w(x) \neq 0, \forall x \in M$ and therefore $M$ is orientable. $\qquad\square$

**Notation**: For $x \in M$ we denote

$$\begin{pmatrix} X^{s,i}(x) \\ X^{s,j}(x) \end{pmatrix} = \begin{pmatrix} X^{1,i}(x) \ldots X^{k,i}(x) \\ X^{1,j}(x) \ldots X^{k,j}(x) \end{pmatrix}$$

and

$$\begin{pmatrix} -X^{s,i}(x) \\ X^{s,j}(x) \end{pmatrix} = \begin{pmatrix} -X^{1,i}(x)X^{2,i}(x) \ldots X^{k,i}(x) \\ X^{1,j}(x)X^{2,j}(x) \ldots X^{k,j}(x) \end{pmatrix}.$$

**Corollary 2.2.** *Let $P$ and $\overline{P}$ be two plane fields in a manifold $M$ such that*
*a)* $TM = P \oplus \overline{P}$, *(i.e.,* $T_x M = P(x) \oplus \overline{P}(x), \forall x \in M$*);*
*b)* $P$ *and* $\overline{P}$ *are orientable.*
*Then, $M$ is orientable.*

**Proof.** Exercise.

**Example 2.5.** Let $\overline{\mathcal{F}}$ be the Reeb foliation in the Moebius band (see Section 1.2.2). Then $T\overline{\mathcal{F}}$ is not orientable. To see this we let $M_\varepsilon = [-1 + \varepsilon, 1 - \varepsilon] \times \mathbb{R}, \quad \varepsilon > 0$ and $M_\varepsilon / F$ be the Möebius band. If $T\overline{\mathcal{F}}$ were orientable, then $T\overline{\mathcal{F}}/(M_\varepsilon/F)$ would be orientable. There is a line field $P$ in $M_\varepsilon/F$ which is orientable. In fact: it suffices to choose $P(x)$ as $T\mathcal{F}_1$ where $\mathcal{F}_1$ is the projection of the vertical foliation in $M_\varepsilon/F$. $P$ is induced by the vertical vector field $X(x, y) = (0, 1)$. Note that $X$ induces a vector field in $M_\varepsilon/F$ since

$$DF(0, 1) = \begin{pmatrix} -1 & 0 \\ 0 & 1 \end{pmatrix} \begin{pmatrix} 0 \\ 1 \end{pmatrix} = \begin{pmatrix} 0 \\ 1 \end{pmatrix},$$

hence $P$ is orientable. But then $T(M_\varepsilon/F) = P \oplus (T\overline{\mathcal{F}}/(M_\varepsilon/F))$ would be orientable, a contradiction. The result follows.

**Definition 2.5.** A foliation $\mathcal{F}$ is *orientable* (resp. *transversely orientable*) if its associated plane field $T\mathcal{F}$ is.

Note that if $M$ supports a foliation $\mathcal{F}$ which is both orientable and transversely orientable, then $M$ is orientable (as a manifold). If $M$ is orientable, then a foliation $\mathcal{F}$ in $M$ is transversely orientable $\Leftrightarrow \mathcal{F}$ is orientable. If $\mathcal{F}$ is a codimension one transversely orientable foliation, then there is a vector field $X$ in $M$ such that $X \pitchfork \mathcal{F}$. Warning: The above does not implies that the time-$t$ map of $X$ preserves $\mathcal{F}$, (*i.e.*, $X_t(\mathcal{F}_x)$ is a leaf for every leaf $\mathcal{F}_x$, $x \in M$).

## 2.3  Orientability of singular foliations

A $C^r$ *singular foliation on a surface* $S$ is a $C^r$-foliation $\mathcal{F}$ in the complement $S \setminus \text{sing}(\mathcal{F})$ of a discrete set $\text{sing}(\mathcal{F})$ in the interior of $S$ which is either transverse or tangent to the boundary of $S$. We denote by $\mathcal{F}_x$ the leaf of $\mathcal{F}$ containing $x \in S \setminus \text{sing}(\mathcal{F})$. One says that $\mathcal{F}$ is $C^r$-*locally orientable* if there is an open cover $\{U_i : i \in I\}$ of $S$ and a $C^r$ vector fields $Y_i$ in $U_i$ such that $\text{sing}(Y_i) = U_i \cap \text{sing}(\mathcal{F})$ and $T_x\mathcal{F}_x = \text{Span}(Y_i(x))$, $\forall x \in U_i \setminus \text{sing}(Y_i)$, where $\text{sing}(Y_i)$ denotes the set of zeroes of $Y_i$. One says that $\mathcal{F}$ is $C^r$ orientable if the cover $\{U_i : i \in I\}$ above can be chosen with a single element $U_1 = S$. This notion of orientability differs from the corresponding one for non-singular foliations due to the presence of the singularities. One can easily construct singular foliations in $D^2$ which are not locally orientable (it suffices to complete the ones described in Figure 2.1 to the whole $D^2$). Clearly a $C^r$ locally orientable singular foliation is $C^r$ orientable. The converse is false in general but true when $S = D^2$, the 2-disc in $\mathbb{R}^2$. Indeed, let $\mathcal{F}$ be a $C^r$ singular foliation in $D^2$. For an open set $U$ of $D^2$ one defines $\mathcal{X}_{\mathcal{F}}^r(U)$ as the space of $C^r$ vector fields in $U$ such that $\text{sing}(Y) = U \cap \text{sing}(\mathcal{F})$ and $T_x\mathcal{F}_x = \text{Span}(Y(x))$, $\forall x \in \setminus \text{sing}(Y)$. A finite family $U_1, \cdots, U_k$ of open sets in $D^2$ is a chain whenever $U_i \cap U_{i+1} \neq \emptyset$ is connected for all $1 \leq i \leq k - 1$. Given a chain $U_1, U_2$ and $Y_1, Y_2 \in \mathcal{X}_{\mathcal{F}}^r(U_1), \mathcal{X}_{\mathcal{F}}^r(U_2)$ we define $\Phi_{U_1,U_2}(Y_1, Y_2)$ to be either $Y_2$ (if $Y_1, Y_2$ have the same orientation in $U_1 \cap U_2$) or $-Y_2$ (otherwise). This definition makes sense because $U_1 \cap U_2 \neq \emptyset$ is connected. Clearly $\Phi_{U_1,U_2}(Y_1, Y_2) \in \mathcal{X}_{\mathcal{F}}^r(U_2)$ and both $Y_1$ and $\Phi_{U_1,U_2}(Y_2)$ have the same orientation in $U_i \cap U_2$. For general chains $U_1, \cdots, U_k$ and $Y_i \in \mathcal{X}_{\mathcal{F}}^r(U_i)$ ($i = 1, \cdots, k$) we define $Z_1 = Y_1$, $Z_{i+1} = \Phi_{U_i,U_{i+1}}(Z_i, Y_{i+1})$ and $\Phi_{U_1,\cdots,U_k}(Y_1, \cdots, Y_k) = Z_k$. Under this definition one has

$$\Phi_{U_1,\cdots,U_k}(Y_1, \cdots, Y_k) = \Phi_{U_{k-1},U_k}(\Phi_{U_1,\cdots,U_{k-1}}(Y_1, \cdots, Y_{k-1}), Y_k). \quad (2.1)$$

Now let us assume that $\mathcal{F}$ is $C^r$ locally orientable and let $\{Y^i \in \mathcal{X}_{\mathcal{F}}^r(U_i) : i = 1, \cdots, r\}$ be a fixed $C^r$ local orientation of $\mathcal{F}$. We can assume that all the $U_i$'s are balls, and so, $U_i \cap U_j$ is either empty or connected for all $i, j$. Define $\tilde{Y}^1 = Y^1$ and for $i = 2, \cdots, r$ we define $\tilde{Y}^i = \Phi_{U_{i_1}, \cdots U_{i_k}}(Y_{i_1}, \cdots, Y_{i_k})$, for some chain $U_{i_1}, \cdots, U_{i_k}$ with $i_1 = 1$ and $i_k = i$. The simply connectedness of $D^2$ implies that the value of $\tilde{Y}_i$ does not depend on the chosen chain $U_{i_1}, \cdots, U_{i_k}$. Let us prove that if $U_i \cap U_j \neq \emptyset$, then $\tilde{Y}_i$ and $\tilde{Y}_j$ have the same orientation in $U_i \cap U_j$. In fact, let $U_{i_1}, \cdots, U_{i_k}$ and $U_{j_1}, \cdots, U_{j_s}$ be two chains realizing $\tilde{Y}_i$ and $\tilde{Y}_j$ respectively. Hence $U_{i_1}, \cdots, U_{i_k}, U_j$ is a chain, and so, the invariance of $\Phi$ with respect to the chains implies

$$\tilde{Y}_j = \Phi_{U_{i_1}, \cdots, U_{i_k}, U_j}(Y_{i_1}, \cdots, Y_{i_k}, Y_j).$$

Then Eq.(2.1) and $i_k = i$ implies

$$\tilde{Y}_j = \Phi_{U_i, U_j}(\Phi_{U_{i_1}, \cdots, U_{i_k}}(Y_{i_1}, \cdots, Y_{i_k}), Y_j) = \Phi_{U_i, U_j}(\tilde{Y}_i, Y_j)$$

proving that $\tilde{Y}_i$ and $\tilde{Y}_j$ have the same orientation in $U_i \cap U_j$ as desired. Next we consider a $C^\infty$ partition of the unity $\{Q_1, \ldots, Q_r\}$ of the cover $U_1, \cdots, U_r$ and define

$$Y = \sum_{i=1}^{k} Q_i \tilde{Y}_i.$$

This vector field yields a $C^r$ orientation of $\mathcal{F}$.

## 2.4 Orientable double cover

Let $P$ be a $k$-plane field in a manifold $M^m$. Define $B_x(M)$ the set of ordered basis of $P(x)$ when $x \in M$. Note that $\{v_1, \ldots, v_k\} \neq \{v_2, v_1, \ldots, v_k\} \in B_x(M)$. Define the following relation in $B_x(M)$:

$$(v_1, \ldots, v_k) \approx_x (w_1, \ldots, w_k) \Leftrightarrow \det \begin{pmatrix} v_s \\ w_s \end{pmatrix} > 0. \text{ Remember that}$$

$$\begin{pmatrix} v_s \\ w_s \end{pmatrix} = \begin{pmatrix} v_1, \ldots, v_k \\ w_1, \ldots, w_k \end{pmatrix}, \quad v_i = \sum_{i<1}^{k} a_{ij} w_j.$$

Note that if $(v_1, \ldots, v_k) \approx_x (w_1, \ldots, w_k)$, then $(-v_1, v_2, \ldots, v_k) \not\approx_x (w_1, \ldots, w_k)$. The relation $\approx_x$ is an equivalence since $\det AB = \det A \cdot \det B$. Define $O_x(M) = B_x(M)/\approx_x$ with projection

$$\pi_x \colon B_x(M) \to O_x(M).$$

Note that

$$O_x(M) = \{O_x, O_x'\}$$

has just two elements. Sometimes we use the notation $-O_x = O'_x$. Finally we define

$$\widetilde{M} = \{(x, O_x) : x \in M, O_x \in O_x(M)\}.$$

At first $\widetilde{M}$ depends on the plane field $P$. When $P(x) = T_xM$, $\forall x \in M$, $\widetilde{M}$ is called *the orientable double cover of M*. In general we call it *the orientable double cover of P*.

**Proposition 2.3.** $\widetilde{M}$ *is a m-dimensional manifold.*

**Proof.** Fix $(x_0, O_{x_0}) \in \widetilde{M}$. By the definition of plane field $P$, there is a neighborhood $U$ of $x_0$ and $k$ smooth vector fields $X^1, \ldots, X^k \colon U \xrightarrow{C^r} TU$ such that

$$P(x) = \mathrm{Span}(X^1(x), \ldots, X^k(x)), \ \forall x \in U.$$

Define $\widetilde{U}$ as:

$$\widetilde{U} = \{(x, \pi_x(X^1(x), \ldots, X^k(x)); x \in U\},$$

if $\pi_{x_0}(X^1(x_0), \cdots, X^k(x_0)) = O_{x_0}$ and

$$\widetilde{U} = \{(x, \pi_x(-X^1(x), \ldots, X^k(x)); x \in U\},$$

if $\pi_{x_0}(X^1(x_0), \cdots, X^k(x_0)) = O'_{x_0}$.

We consider the projection

$$\pi \colon \widetilde{M} \to M; \ \pi(x, O_x) = x.$$

We can assume that $U$ is the domain of a local chart $(U, Q)$ around $x_0$. Define the chart $(\widetilde{U}, \widetilde{Q})$ by $\widetilde{Q} = Q \circ \pi$.

The family $\{(\widetilde{U}, \widetilde{Q})\}$ is an atlas of $\widetilde{M}$. In fact, let $(U, Q)$, $(V, \psi)$ be two such charts with $U \cap V \neq \emptyset$. Then $\widetilde{\psi} \circ \widetilde{Q}^{-1} \colon \widetilde{Q}(\widetilde{U} \cap \widetilde{V}) \to \widetilde{\psi}(\widetilde{U} \cap \widetilde{V})$ satisfies $\widetilde{\psi} \circ \widetilde{Q}^{-1}(y) = \psi \circ Q^{-1}(y)$. Hence $\widetilde{M}$ is a manifold of class $C^r$ and dimension $m$. Moreover, $\pi \colon \widetilde{M} \to M$ is differentiable and even a local diffeomorphism. Note that $T_{(x,O_x)}\widetilde{M} \simeq T_xM$. Define also a $k$-plane field $\widetilde{P}$ in $\widetilde{M}$ given by

$$\widetilde{P}(x, O_2) = a(\pi/\widetilde{U})^{-1}(x) \ (P(x)).$$

$\widetilde{M} = (\widetilde{M}, \pi, \widetilde{P})$ is the orientable double cover of $P$. $\qquad \square$

**Theorem 2.3.** *Let $M$ be a manifold and $P$ a plane field in $M$ with orientable double cover $(\widetilde{M}, \widetilde{P})$. Then, $\widetilde{M}$ is connected if, and only if, $P$ is not orientable.*

**Proof.** Assume that $\widetilde{M}$ is connected. Then we can fix a curve $\widetilde{c} \subset \widetilde{M}$ joining $(x_0, O_{x_0})$ with $(x_0, O'_{x_0})$ The curve $c = \pi \circ \widetilde{c}$ in $M$ is closed and contains $x_0$. Suppose by contradiction that $P$ is orientable. Let $\{U_i\}$ the cover of $M$ and $\{X^{1,i}, \cdots, X^{k,i}\}$ be the vector fields in $U_i$ generating $P$ such that

$$det \begin{pmatrix} X^{s,i}(x) \\ X^{s,j}(x) \end{pmatrix} > 0. \tag{2.2}$$

Because $c$ is compact we can suppose that $U_1, \cdots, U_r$ is a cover of $c$. We can further suppose that $x_0 \in U_1$ and $O_{x_0} = \pi_{x_0}(X^{1,1}(x_0), \cdots, X^{k,1}(x_0))$. Define a new curve $\hat{c} \subset \widetilde{W}$ given by

$$\hat{c}(t) = (c(t), \pi_{c_t}(X^{1,i}(c(t)), \cdots, X^{k,i}(c(t)))), \text{ if } c(t) \in U_i,$$

and $i = 1, \cdots, k$.

Note that $\hat{c}$ is well defined by Eq.(2.2). In addition, $\hat{c}$ is continuous because both $c(t)$ and $t \to \pi_{c(t)}(X^{1,i}(c(t)), \cdots, X^{k,i}(x_0))$ are. Define

$$B = \{t \in [0,1] : \hat{c}(t) = \widetilde{c}(t)\}.$$

We have that $B \neq \emptyset$ because $x_0 \in B$. Moreover, $B$ is closed because $\hat{c}$ and $\widetilde{c}$ are continuous. Let us prove that $B$ is open. In fact, if $t_0 \in B$ then $c(t_0) \in U_{i_0}$ for some $i_0$. If $t_0 \notin \text{Int}(B)$ then there is a sequence $t_n \to t_0$ in $[0,1]$ such that $\hat{c}(t_n) \neq \widetilde{c}(t_n)$ for all $n$. Because $c$ is continuous and $c(t-0) \in U_{i_0}$ we can suppose that $c(t_n) \in U_{i_0}$ for all $n$ yielding

$$\hat{c}(t_n) = (c(t_n), \pi_{c(t_n)}((X^{1,i_0}(c(t_n)), \cdots, X^{k,i_0}(c(t_n)))).$$

Write $\hat{c}(t) = (c(t), \gamma(t))$, where $\gamma(t) \in a(c(t))$ is continuous. Because $\hat{c}(t_n) = \widetilde{c}(t_n)$ one has

$$\pi_{c(t_n)}((X^{1,i_0}(c(t_n)), \cdots, X^{k,i_0}(c(t_n)))) = -\gamma(t_n).$$

By taking limits the last expression yields

$$\pi_{c(t_0)}(X^{1,i_0}(c(t_0)), \cdots, X^{k,i_0}(c(t_0))) = -\gamma(t_0)$$

contradicting $\hat{c}(t_0) = \widetilde{c}(t_0)$. This contradiction shows that $\hat{c}(t) = \widetilde{c}(t)$ for all $t$ and then $\hat{c}(1) = \widetilde{c}(1)$. This would imply $O_{x_0} = -O_{x_0}$ which is absurd. This proves the first part. For the converse, we shall prove that if $\widetilde{M}$ not connected then $P$ is not orientable. Recall the projection $\pi \colon \widetilde{M} \to M$ given by $\pi(x, O_x) = x$. Let $\widetilde{M}'$ be a connected component of $\widetilde{M}$. Observe that $\pi(\widetilde{M}') = M$. In fact, since $\pi$ is a local diffeomorphism we have that $\pi(\widetilde{M}')$ is open in $M$. Let us prove that $\pi(\widetilde{M})$

is closed in $M$. Choose $x_n \in \pi(\widetilde{M'}) \to x \in M$. By definition there is a neighborhood $U$ of $x$ and $k$ vector fields $X^1, \cdots, X^k$ generating $P$ in $U$. Obviously there is $\widetilde{y}'_n \in \widetilde{M}'$ such that $\pi(\widetilde{y}'_n) = x_n$. Note that $\widetilde{y}'_n = (x_n, O_{x_n})$ for some $O_{x_n} \in a(x_n)$. Without loss of generality we can assume that $O_{x_n} = \pi_{x_n}(X^1(x_n), \cdots, X^k, (x_n))$ for all $n$. Passing to the limit the last expression yields $O_{x_n} \to \pi_x(X^1(x), \cdots, X^k, (x)) = O_x$. Hence $\widetilde{y}'_n \to (x, O_x)$. Since $\widetilde{M}'$ is closed in $\widetilde{M}$ we conclude that $(x, O_x) \in \widetilde{M}'$. So, $x = \pi(x, O_x) \in \pi(\widetilde{M'})$ proving that $\pi(\widetilde{M'})$ is closed. Because $M$ is connected we conclude that $\pi(\widetilde{M'}) = M$ as desired.

On the other hand, since $\pi^{-1}(x)$ has two elements for all $x \in M$ we conclude that $\widetilde{M}$ has two connected components which we denote by $\widetilde{M}_1, \widetilde{M}_2$. This implies that $\forall x \in M$ and $\forall O_x \in O(x)$ if $(x, O_x) \in \widetilde{M}_1 \Leftrightarrow (x, -O_x) \in \widetilde{M}_2)$. It follows that $\pi/\widetilde{M}_1$ is one-to-one. Hence $\pi : \widetilde{M}_1 \to M$ is a diffeomorphism. Since $\widetilde{P}$ is orientable, and $P = \pi^{-1}(P)$ we would have that $P$ is orientable, a contradiction. This proves the theorem. □

**Corollary 2.3.** *Every plane field on a simply connected manifold is orientable and transversely orientable. In particular, all simply connected manifolds are orientable.*

**Proof.** Let $P$ be a plane field on a simply connected manifold $M$. If $P$ were not orientable then its double cover $\pi : \widetilde{M} \to M$ is connected. Being $M$ simply connected we would have $M = \hat{M}$, where $\hat{M} \to M$ is the universal cover of $M$. This would imply that $\widetilde{M} \to \hat{M}$ is non-trivial cover, a contradiction. This contradiction proves that $P$ is orientable. That $P$ is transversely orientable follows applying the previous result to a complementary plane field of $P$. The result follows. □

**Example 2.6.** The Reeb foliation in $S^3$ is orientable and transversely orientable (because $\pi_1(S^3) = 1$).

Exercise 2.4.1. Show that any foliation in the solid torus $D^2 \times S^1$ tangent to the boundary is orientable.

## 2.5   Foliations and differentiable forms

Remember that a differential $k$-form $w$ of $M$ is a multilinear map associating to each point $p \in M$ a linear $k$-form in $T_pM$, that is is, $w(p) \in \Lambda^k(T_pM)$, where $\Lambda^k(E)$ denotes the space of $k$-forms in a vector space $E$. The space of all $k$-forms in $M$ is denoted by $\Lambda^k(M)$. If $w \in \Lambda^k(M)$ and $\eta \in \Lambda^\ell(M)$, then the alternating product $w \wedge \eta \in \Lambda^{k+\ell}(M)$. If $w$ is a $k$-form, then there is a derivative $dw$ of $w$, $\quad d = d_k \colon \Lambda^k(M) \to \Lambda^{k+1}(M)$. The form $w$ is *closed* if $dw = 0$, and $w$ is *exact* if $w = d\eta$. As it is well-known:

**Poincaré's lemma**: *An exact form of class $C^2$ is closed, i.e., $\quad d(dw) = d^2(w) = 0$, $\forall w \in \Lambda^k(M)$. Conversely, a $C^1$ closed form is locally exact.*

Denote by $Z^k(M) = \mathrm{Ker}(d_k)$ is the set of closed $k$-forms and by $B^k(M) = d_{k-1}(\Lambda^{k-1}(M))$ is the set of exact $k$-forms. Poincaré's lemma implies that exact forms are closed, namely $B^k(M) \subset Z^k(M)$. The quotient space $H^k(M) = Z^k(M)/B^k(M)$ is called the (de Rham) cohomology $k$-group of $M$.

Let $w \in \Lambda^1(M^n)$ be a non-singular 1-form namely, $w(p) \neq 0$, $\forall p \in M$. The map $p \mapsto \mathrm{Ker}(w(p))$ defines a $(n-1)$-plane field in $M$.

We leave the following proposition as an exercise for the reader (use Frobenius' theorem).

**Proposition 2.4.** *The following conditions are equivalent for a smooth differentiable one-form $w \in \Lambda^1(M)$.*

*(1) $w$ is integrable if.*
*(2) $dw \wedge w = 0$.*
*(3) $dw = w \wedge \eta$ for some $\eta \in \Lambda^1(M)$.*

**Example 2.7.** If $w$ is closed, then $w$ is integrable and so $w$ defines a foliation $F_w$. This remark apply to the following. Define $M = \mathbb{R}^2$, $w = adx + bdy$, $a, b \in \mathbb{R}$. Then $w$ is closed $\Rightarrow w$ induces a foliation $F_w$. The leaves of this foliation are given by the solution of the differential equation

$$adx + bdy = 0 \Rightarrow y' = -b/a.$$

The general solution of this equation is the straight-line family $y = (-b/a) \cdot x + K$, $K \in \mathbb{R}$. This gives a foliation of $\mathbb{R}^2$ by these straight-lines.

**Example 2.8 (Thurston).** Let $L$ be a closed manifold with $H^1(L) \neq 0$. Let $\alpha$ be a closed non-exact 1-form of $L$ and $f \colon S^1 \to \mathbb{R}$ be a differentiable map. Denote by $d\theta$ the standard 1-form of $S^1$. We define the 1-form $w$ in the product $M = L \times S^1$ given by

$$w = d\theta + f(\theta)\alpha.$$

Note that

$$dw = d(d\theta + f(\theta)\alpha) = d(f(\theta)\alpha) = f'(\theta) \cdot d\theta \wedge \alpha + f(\theta) \cdot d\alpha$$

and $d\alpha = 0$. It follows that

$$dw \wedge w = (f'(\theta) \cdot d\theta \wedge \alpha) \wedge (d\theta + f(\theta)\alpha)$$

$$= f'(\theta) \cdot d\theta \wedge \alpha \wedge d\theta + f'(\theta) \cdot d\theta \wedge \alpha \wedge f(\theta) \cdot \alpha = 0.$$

It follows that $w$ is integrable, *i.e.*, $Ker(w)$ is tangent to a $C^1$ codimension one foliation $\mathcal{F}_w$ in $M$. Note that the sets $L_{\theta_0} = \{(x, \theta) \in M : \theta = \theta_0\}$, where $\theta_0 \in f^{-1}(0)$ are compact leaves of $\mathcal{F}_w$. In fact, fix $\theta_0 \in f^{-1}(0)$ and $(x, \theta) \in L_{\theta_0}$. If $v_{(x,\theta)} \in T_{(x,\theta)}L_{\theta_0} \Rightarrow v_{(x,\theta)} = (v_x, 0)$ and $\theta = \theta_0$. Hence

$$w(v_{(x,\theta)}) = w(v_x, 0) = d\theta(0) + f(\theta_0)\alpha(v_x) = 0 + 0 \cdot \alpha(v_x) = 0$$

proving that $L_{\theta_0}$ is a leaf of $\lambda_w$. Clearly $L_{\theta_0}$ is diffeomorphic to $L$ and so $L_{\theta_0}$ is a compact leaf of $\mathcal{F}_w$ (recall that $L$ is closed).

# Chapter 3

# Topology of the leaves

## 3.1 Space of leaves

Let $\mathcal{F}$ be a foliation of a manifold $M$. Given $z \in M$ we denote by $\mathcal{F}_z$ the leaf of $\mathcal{F}$ that contains $z$. The relation $x, y \in M$, $x \sim y \Leftrightarrow x \in \mathcal{F}_y \Leftrightarrow y \in \mathcal{F}_x$ is an equivalence. The quotient space $O_{\mathcal{F}} = M/N$ is called the *space of leaves of $\mathcal{F}$*. Denote by $\pi \colon M \to O_{\mathcal{F}}$ the projection. We set in $O_{\mathcal{F}}$ the topology making $\pi$ continuous, namely $V \subseteq O_{\mathcal{F}}$ is open $\Leftrightarrow \pi^{-1}(V) \subset M$ is. If $A \subseteq M$ we define $\mathcal{F}(A) = \text{Sat}(A) = \bigcup_{x \in A} \mathcal{F}_x$ and call it the *saturation of $A$*. This set is formed by those $x$ such that $\mathcal{F}_x$ meets $A$. The set $A$ is *saturated* by $\mathcal{F}$ if $\mathcal{F}(A) = A$.

**Example 3.1.** The leaf space of the foliation $F_2$ in Figure 1.2 is not Hausdorff. In fact the vertical boundaries of $I \times \mathbb{R}$ correspond to elements in the leaf space which cannot be separated by open sets.

**Proposition 3.1.** *The saturation $\mathcal{F}(A) \subset M$ of an open set $A \subset M$ is open.*

**Proof.** Choose $x \in \mathcal{F}(A)$. By definition, $\mathcal{F}_x \cap A \neq \emptyset$, hence there is $y \in \mathcal{F}_x \cap A$. There there exists a finite collection of plaques $\alpha_1, \alpha_2, \ldots, \alpha_k$ of $\mathcal{F}$ in $\mathcal{F}_x$ such that $\alpha_i \cap \alpha_{i+1} \neq \emptyset$, $x \in \alpha_k$ $y \in \alpha_1$. Let $U_i$ be the domain of the chart of $\mathcal{F}$ defining $\alpha_i$. Because $y \in A$, $A$ is open, we can suppose $U_i \subseteq A$. We project $U_i$ into an open set of $\mathcal{F}(A)$ containing $x$ as follows: consider $U_1 \cap U_2$ which is non-empty since $\emptyset \neq \alpha_1 \cap \alpha_2 \subset U_1 \cap U_2$. Let $\tilde{U}_1 = U_1$, $\tilde{U}_2 = U\{\alpha; \alpha$ be a plaque of $U_2$ with $\alpha \cap \tilde{U} \neq \emptyset\}$. Note that $\alpha_2 \subset \tilde{U}_2$. We have $\tilde{U}_2$ is open and $\tilde{U}_2 \subseteq \mathcal{F}(A)$ (since, for a plaque $\alpha$ we have $\alpha \cap \tilde{U}_2 \neq \emptyset \Rightarrow \alpha \cap \tilde{U}_1 \neq \emptyset \Rightarrow \alpha \cap \mathcal{F}(A) \neq \emptyset \Rightarrow \alpha \subset \mathcal{F}(a))$. Also, $\alpha_2 \cap \alpha_3 \neq \emptyset$. Define $\tilde{U}_3 = U\{\alpha, \alpha$ is a plaque of $U_3$ with $\alpha \cap \tilde{U}_2 \neq \emptyset\}$. As

$\alpha_2 \cap \alpha_3 \neq \emptyset$ and $\alpha_2 \subset \widetilde{U}_2$, we have that $\alpha_1 \subset \widetilde{U}_3$. Hence $\widetilde{U}_3$ is open and $\widetilde{U}_3 \subseteq \mathcal{F}(A)$. Inductively we have $\widetilde{U}_i$, $\forall i = 1, \ldots, k$, such that $\alpha_i \subset \widetilde{U}_i$, $\widetilde{U}_i$ is open and $\widetilde{U}_i \subset \mathcal{F}(A)$. Hence $x \in \alpha_k \subset \widetilde{U}_k$ and $\mathcal{F}(A)$ is open. $\qquad \square$

**Corollary 3.1.** *The projection* $\pi \colon M \to O_{\mathcal{F}}$ *is open, (i.e., it sends open sets into open sets)*

**Proof.** Let $A \subset M$ be open. We have

$$x \in \mathcal{F}(A) \Leftrightarrow \mathcal{F}_x \cap A \neq \emptyset \Leftrightarrow \exists y \in \mathcal{F}_x \cap A \Leftrightarrow \mathcal{F}_x = \mathcal{F}_y$$

and

$$y \in A \Leftrightarrow \pi(x) = \pi(y) \in \pi(A) \Leftrightarrow \pi(x) \in \pi(A) \Leftrightarrow x \in \pi^{-1}(\pi(A)).$$

Therefore

$$\mathcal{F}(A) = \pi^{-1}(\pi(A)).$$

Because $\mathcal{F}(A)$ is open we have that $\pi(A)$ is open with respect to the quotient topology. $\qquad \square$

**Warning**: Not every projection is open. For instance, consider the projection of the parabola $y = x^2$, in $\mathbb{R}^2$, into the $y$-axis.

**Definition 3.1.** We say that a subset $A \subset M$ is *invariant* for $\mathcal{F}$ (or $\mathcal{F}$-*invariant*) if $A = \mathcal{F}(A) = \mathrm{Sat}(A)$.

**Lemma 3.1.** *If $A$ is $\mathcal{F}$-invariant then $\partial A, \mathrm{Int}(A)$ and $\overline{A}$ are $\mathcal{F}$-invariant.*

**Proof.** We have $\mathrm{Int}(A)$ is open. Thus $\mathcal{F}(\mathrm{Int}(A))$ is open and therefore $\mathrm{Int}(A) \subseteq \mathcal{F}(\mathrm{Int}(A)) \subset \mathcal{F}(A) = A$. Since $\mathrm{Int}(A)$ is the biggest open set contained in $A$ we have $\mathcal{F}(\mathrm{Int}(A)) = \mathrm{Int}(A)$. Assume now that $A$ is $\mathcal{F}$-invariant. Then $M \backslash A$ is also invariant. Thus $\mathrm{Int}(M \backslash A)$ is invariant and $\mathrm{Int}(M \backslash A) = M \backslash \overline{A} \Rightarrow M \backslash \overline{A}$ is $\mathcal{F}$-invariant. Therefore, $\overline{A}$ is $\mathcal{F}$-invariant. Finally, $\partial A = \overline{A} \backslash \mathrm{Int}(A)$ where $\overline{A}$ and $\mathrm{Int}(A)$ are $\mathcal{F}$-invariant. Hence $\partial A$ is $\mathcal{F}$-invariant. $\qquad \square$

**Theorem 3.1.** *Let $F$ be a leaf of a foliation $\mathcal{F}$ and $\Sigma$ be a transverse section of $\mathcal{F}$ intersecting $F$. Then, one of the following alternatives holds:*

*(1) $F \cap \Sigma$ is discrete.*
*(2) $\overline{F \cap \Sigma}$ has non-empty interior in $\Sigma$.*

*(3)* $\overline{F \cap \Sigma}$ *is a perfect set, i.e., a closed set without isolated points with empty interior.*

**Proof.** It suffices to prove that if $F \cap \Sigma$ is not discrete then $\overline{F \cap \Sigma}$ is perfect. Suppose by contradiction that $\overline{F \cap \Sigma}$ is not perfect, *i.e.*, $\overline{F \cap \Sigma}$ has an isolated point $x_0$. Because $x_0$ is isolated in $\overline{F \cap \Sigma}$ we have $x_0 \in F$. Because $F \cap \Sigma$ is not discrete, there is $x^* \in F \cap \Sigma$ which is an accumulation point of $\{x_n\} \subset F \cap \Sigma$, $x_n \neq x^*$. Because $x_n \in F$ we have that $F$ passes arbitrarily close to $x^*$. Using a suitable plaque sequence we can see that $F$ passes close to $x_0$ (see Figure 3.1). This is a contradiction and the proof follows. □

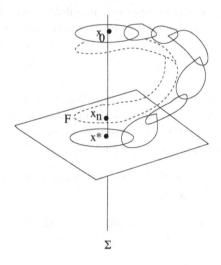

Fig. 3.1

Exercise 3.1.1. Let $\mathcal{F}$ be a foliation on $M$ of codimension $q$. A differentiable map $f \colon N \to M$ is *transverse* to $\mathcal{F}$ if it is transverse to each leaf $L \in \mathcal{F}$ as an immersed submanifold in $M$. Show that in this case there is a naturally defined foliation $f^*(\mathcal{F})$ in $N$ of codimension $q$ such that for each leaf $L \in \mathcal{F}$ the inverse image $f^{-1}(L)$ is a union of leaves of $f^*(\mathcal{F})$.

## 3.2 Minimal sets

Let $\mathcal{F}$ be a foliation in $M$. A subset $\mu \subseteq M$ is called *minimal* for $\mathcal{F}$ if

1) $\mu$ is closed and $\mathcal{F}$-invariant.

2) if $\emptyset \neq \mu' \subseteq \mu$ satisfies (1), then $\mu' = \mu$.

**Remark 3.1.** Zorn's lemma applied to the set of closed invariant subsets of $M$ (ordered by inclusion) implies that there is at least one minimal subset. Every closed leave is a minimal set. If $\mu$ is minimal and $F$ is a leaf of $\mathcal{F}$ in $\mu$, then $\overline{F} = \mu$. If $\mu$ is minimal and meets a closed leaf $F$, then $\mu = F$. In general the set of minimal sets is pairwise disjoint. The Reeb foliation in $S^3$ has a unique minimal set which is its compact toral leaf. The irrational foliation in $T^2$ has $T^2$ as its unique minimal set. As we shall see later on there is no minimal foliation, (*i.e.*, the whole manifold is minimal) in $S^3$ of codimension 1 (by Novikov's compact leaf theorem). Analogously there are no minimal foliations on compact manifolds with finite fundamental group. Any foliation in the Klein bottle has a compact leaf. Hence there is no minimal foliation in the Klein bottle. A foliation is *transitive* if it has a dense leaf. Minimal foliations are transitive. The converse however, is not true. There is no transitive codimension one foliations on compact 3-manifolds with finite fundamental group. In particular $S^3$ does not support transitive codimension one foliations. Minimal foliations have no compact leaves. As we shall see later, a transitive codimension one foliation on a compact 3-manifold, has no compact leaves as well.

**Exercise 3.2.1.** Find a non-minimal transitive codimension one foliation on a compact three-manifold $M^3$.

**Lemma 3.2.** *Suppose that $\mathcal{F}$ is a foliation in $M$ and $\mu$ is minimal for $\mathcal{F}$. Then,* $\mathrm{Int}(\mu) \neq \emptyset \Leftrightarrow \mu = M$.

**Proof.** Clearly $\mu = M \Rightarrow \mathrm{Int}(\mu) \neq \emptyset$.

Conversely, let $\mu$ be minimal with $\mathrm{Int}(\mu) \neq \emptyset$. On one hand $\mu$ is closed by definition. On the other hand, $\mu$ is open since $\mu = \mathrm{Int}(\mu)$ as $x \in \mu \Rightarrow \overline{\mathcal{F}}_x = \mu \Rightarrow \overline{\mathcal{F}}_x \cap \mathrm{Int}(\mu) \neq \emptyset \Rightarrow x \in \mathcal{F}(\mathrm{Int}(\mu)) = \mathrm{Int}(\mu))$. Since $M$ is connected we conclude that $\mu = M$ and the proof follows. $\square$

**Proposition 3.2.** *Suppose that $\Sigma$ is a p-disc (p $=\mathrm{cod}\mathcal{F}$), $\mu$ is a minimal subset of $\mathcal{F}$ with $\mu \cap \Sigma \neq \emptyset$ and $\mu \cap \partial\Sigma = \emptyset$. If $\mu$ is not a closed leaf then $\mu \cap \Sigma$ is perfect.*

**Proof.** Assume that $\mu$ is not a closed leaf and prove that $\mu \cap \Sigma$ is a perfect set. For this we proceed as follows. Observe that $\Sigma$ is compact since it is a $p$-disc. Let $F \subset \mu$ be a leaf of $\mathcal{F}$. It suffices to prove that $\overline{F \cap \Sigma}$ is perfect. By contradiction suppose that it is not so. Then either $F \cap \Sigma$ is discrete or $\overline{F \cap \Sigma}$ has non-empty interior by Theorem 3.1. In the later case $\mu \cap \Sigma$ has non-empty interior (in $\Sigma$) since $F$ is dense and $\Sigma$ is compact. This would imply that $\mu$ has non-empty interior and then $\mu = M$ by the previous lemma. This is a contradiction because $\mu \cap \partial \Sigma = \emptyset$. We conclude that $F \cap \Sigma$ is discrete, and so, $F \cap \Sigma$ is finite. If $F$ were not closed then we could find $x \in \overline{F} \setminus F$. Because $F \subset \mu$ and $x \in \overline{F}$ we have $x \in \mu$. Because $\mu$ is minimal the leaf $F_x$ of $\mathcal{F}$ containing $x$ is dense in $\mu$. In particular $F_x \cap \text{Int}(\Sigma) \neq \emptyset$ (recall $\mu \cap \partial \Sigma = \emptyset$). By applying the argument described in Figure 3.1, we would have that $F$ intersects $\Sigma$ infinitely many times, a contradiction. This contradiction proves that $F$ is a closed leaf. Since $F$ is dense in $\mu$ we would have that $\mu = F$ is a closed leaf which is impossible. This contradiction proves the result. $\qquad\square$

**Definition 3.2.** A minimal set of a foliation on a manifold $M$ is *exceptional* if it is neither a closed leaf nor the whole manifold $M$.

A very interesting and deep problem is:

**Problem 3.1.** *Find necessary and sufficient conditions for the existence of exceptional minimal sets.*

**Lemma 3.3.**

*In codimension one minimal sets are described below. This is a straightforward consequence of the above results and of the well-known description of perfect sets in the real line. A nowhere dense minimal set of a codimension one foliation is either a closed leaf or an exceptional minimal set. In particular, an exceptional minimal set of a codimension one foliation, meets a transverse section in a set homeomorphic to a Cantor set.*

**Definition 3.3.** A leaf $L$ of a codimension one foliation $\mathcal{F}$ on a manifold $M$ is called *exceptional* if for some (and therefore for any) transverse section $\Sigma \subset M$ to $\mathcal{F}$ meeting $L$ the intersection $\Sigma \cap L$ is nowhere dense and without isolated points. In this case, the closure $\overline{\Sigma \cap L \subset \Sigma \hookrightarrow \mathbb{R}}$, is homeomorphic to a Cantor set.

**Example 3.2.** The irrational foliation in $T^2$ is minimal, and so, it has no exceptional minimal sets. In fact a foliation arising from a $C^2$ vector field on a closed surface has no exceptional minimal sets. This is false for $C^1$ vector fields by the classical Denjoy's counterexample. The Reeb foliation in $S^3$ has no exceptional minimal sets.

**Example 3.3.** Let $B$ be the bitorus and consider the representation $Q$ : $\pi_1(B) \to \text{Diff}(S^1)$ as described in Section 1.2.6. Recall that the behavior of the suspended foliation $\mathcal{F}_Q$ depends on the maps $f, g \in \text{Diff}(S^1)$ used in the construction of $Q$. By a suitable choice of $f, g$ we have that $\mathcal{F}_Q$ is a $C^\infty$ foliation in some closed Seifert 3-manifold exhibiting an exceptional minimal set. This example (due to Sacksteder) gives a counterexample for a possible version of the Denjoy's theorem.

Exercise 3.2.2. Are there transitive codimension one foliations with exceptional minimal sets?

# Chapter 4

# Holonomy and stability

## 4.1 Definition and examples

An important tool for the study of foliations is the notion of *holonomy group* defined as follows. Let $\mathcal{F}$ be a foliation on a manifold $M$. Let $U_i, U_j$ be two charts of $\mathcal{F}$ with $U_i \cap U_j \neq \emptyset$. Denote by $\pi_i : U_i \to \Sigma_i$ and $\pi_j : U_j \to \Sigma_j$ the projection along the plaques. Suppose that every plaque (of $\mathcal{F}$) in $U_i$ intersects at most one plaque in $U_j$. Then we can define

$$f_{i,j}(x) = \pi_j(\alpha_x),$$

where $x \in \Sigma_i$ and $\alpha_x$ is the unique plaque of $U_i$ containing $x \in U_i$. The resulting map

$$f_{i,j} : \mathrm{Dom}(f_{i,j}) \subset \Sigma_i \to \Sigma_j$$

is called the *holonomy map* induced by the two foliated charts $(U_i, X_i), (U_j, X_j)$. Let $U_1, \cdots, U_r$ be a finite family of foliated charts such that every plaque of $U_i$ intersects at most one plaque of $U_j$ (for all $i, j$). We can define the *holonomy map* $f_{1,\cdots,r} : \mathrm{Dom}(f_{1,\cdots,r}) \subset \Sigma_1 \to \Sigma_r$ by

$$f_{1,\cdots,r} = f_{r-1,r} \circ f_{r-2,r-1} \circ \cdots \circ f_{1,2}.$$

Now, let $L$ be a leaf of $\mathcal{F}$ and $x, y \in L$. Clearly $L$ is connected (by definition) and so there is a curve $c : [0, 1] \to L$ joining $x$ and $y$. This curve can covered by a finite family of foliated charts $U_1, \cdots, U_r$ such that $x \in U_1, y \in U_r$ and every plaque of $U_i$ intersects at most one plaque of $U_j$ (for all $i, j$). Without loss of generality we can assume that $x \in \Sigma_1, y \in \Sigma_r$. The map

$$f_c = f_{1,\cdots,r}$$

is the holonomy induced by the curve $c$. Note that by definition we have $f_c(x) = y$. One can easily prove that $f_c$ does not depend on the foliated cover $U_1, \cdots, U_r$. Moreover, $f_c$ only depends on the homotopy class of $c$. More precisely, if $c, c' \subset L$ are homotopic in $L$ (with fixed end points) then $f_c = f_{c'}$ in an open subset of $\Sigma_1$ containing $x$. When $x = y$ we obtain a representation

$$\Phi : \pi_1(L) \to \mathrm{Germ}(\Sigma)$$

given by

$$\Phi(\gamma) = [f_c],$$

where $c$ is a representative of $\gamma \in \pi_1(L)$, $\Sigma$ is a transverse of $\mathcal{F}$ containing $x \in L$ ($\Sigma \approx \Sigma_1$) and

$$\mathrm{Germ}(\Sigma) = \{f \colon \mathrm{Dom}(f) \subset \Sigma \to \Sigma : f(x) = x\}/\approx$$

is the group of germs of $C^r$ maps with fixed at $x$. The equivalence relation $\approx$ on the space of $C^r$ maps $\{f \colon \mathrm{Dom}(f) \subset \Sigma \to \Sigma : f(x) = x\}$ is defined by: $f \approx g$ if and only if $f$ and $g$ coincide in a neighborhood of $x$.

**Definition 4.1.** The image $\mathrm{Hol}(L, \Sigma, x) = \Phi(\pi_1(L))$ of $\Phi$ is called the *holonomy group of $L$* with respect to $\Sigma$ and $x$.

Up to conjugacy by $C^r$ germs of diffeomorphisms, the group $\mathrm{Hol}(L, \Sigma, x)$ does not depend on $\Sigma$ and $x \in L$. This allows us to define the *holonomy group of the leaf $L$* as the conjugacy class $\mathrm{Hol}(\mathcal{F}, L) = \mathrm{Hol}(L)$ of the groups $\mathrm{Hol}(L, \Sigma, x)$, under conjugation by $C^r$ germs of diffeomorphisms. A representative of $\mathrm{Hol}(L)$ is then a group $\mathrm{Hol}(L, \Sigma, x)$. The algebraic and dynamical properties of the holonomy group $\mathrm{Hol}(L)$ can be read on any representative $\mathrm{Hol}(L, \Sigma, x)$ as we shall see.

We say that the leaf $L$ has *no holonomy* or *is without holonomy* or still *has trivial holonomy* if $\mathrm{Hol}(L) = \{\mathrm{Id}\}$. A foliation *without holonomy* is a foliation whose leaves are without holonomy.

**Example 4.1.** A simply connected leaf has trivial holonomy. In particular a foliation by planes $\mathbb{R}^2$ is without holonomy.

**Example 4.2.** There are foliations admitting non-simply connected leaves without holonomy. Indeed, define $M_0 = I \times T^2$ where $I$ is a compact interval. Then $M_0$ has a boundary formed by two torii $T_1$ (external one) and $T_2$ (internal one), see Figure 4.1. Gluing $T_1$ with $T_2$ by a diffeomorphism $\varphi : T_1 \to T_2$, we obtain a closed manifold $M$. The trivial foliation of $M_0$ formed by concentric torii $* \times T^2$, $* \in I$ defines a foliation $\mathcal{F}$ of $M$. Any

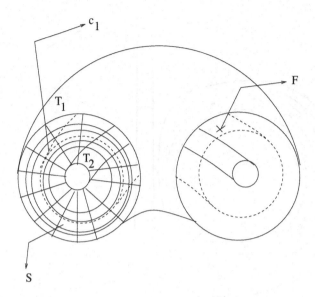

Fig. 4.1

leaf of $\mathcal{F}$ is a torus. Every torus bundles over $S^1$ can be obtained in this way. We observe that $\mathcal{F}$ is a foliation without holonomy. This can be seen as follows. Let $F$ be a leaf of $\mathcal{F}$. Then $\pi_1(F) = \mathbb{Z}^2$ is the free abelian group with two generators $[c_1], [c_2]$, where $c_1, c_2$ are the meridian curve and the parallel curve in $T^2$ respectively. The generator $c_1$ is depicted at Figure 4.1. Consider the transverse surface $S = I \times c_1$ in Figure 4.1. Note that $\mathcal{F}$ intersect $S$ in a circle foliation. The holonomy induced by $c_1$ in $S$ is precisely the first return induced by this circle foliation. Since this return map is the identity one has $\Phi([c_1]) = \text{Id}$. A similar argument shows that $\Phi([c_2]) = \text{Id}$. Since $[c_1], [c_2]$ are the generators of $\pi_1(F)$ we conclude that $\text{Hol}(F) = \Phi(\pi_1(F)) = 0$. This proves that $\mathcal{F}$ has no holonomy as desired. With similar arguments we can prove that all foliation arising from a surface bundle over $S^1$ are without holonomy.

**Example 4.3 (Holonomy of the Reeb foliation).** Let $\mathcal{F}$ be the Reeb foliation in $S^3$ described in Chapter 1 Example 1.3. Then $\mathcal{F}$ has a torus leaf $T$ and all remaining leaves are planes (and so they have in the holonomy). To calculate $\text{Hol}(T)$ we proceed as in the previous example. Indeed, as before $\pi_1(T)$ is generated by the meridian curve and the parallel curve $c_1, c_2$.

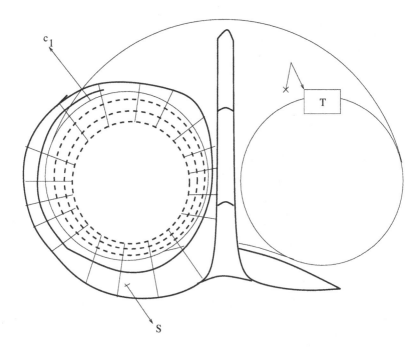

Fig. 4.2

If $S$ is a transverse annulus as in Figure 4.2, $\Sigma$ is a transverse interval in $S$ centered at $x_0 \in \Sigma \cap T$ then the foliation induces a flow on it whose return map $f$ is as in the right-hand side figure at Figure 4.3.

This map is precisely the holonomy of $c_2$. Analogously $c_2$ produces a holonomy having the graph depicted in the left-hand side figure at Figure 4.3. Now $\mathrm{Hol}(T)$ is generated by the classes of there two maps. Note that $\mathrm{Hol}(T)$ is abelian since it is the homomorphic image of $\mathbb{Z}^2$ (which is abelian). Because $\mathrm{Hol}(T)$ is torsion free we conclude that $\mathrm{Hol}(T) = \mathbb{Z}^2$.

**Example 4.4.** A foliation $\mathcal{F}$ tangent to a closed non-singular $C^\infty$ 1-form $w$ in a manifold $M$ has trivial holonomy. Indeed, let $X$ be the gradient of $w$ defined by

$$w_p(v_p) = \langle X(p), v_p \rangle,$$

for all $p \in M$ and $v_p \in T_pM$. Clearly $X$ is non-singular since $w$ is. In addition $\mathcal{F}$ is transverse to $\mathcal{F}$. Let $F$ be a leaf of $\mathcal{F}$ and $c$ a closed curve in $F$. We can assume that $c : S^1 \to F$ is an immersion. Set $I = [-1, 1]$ and

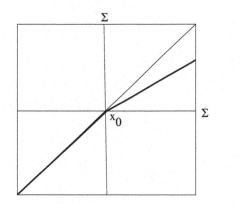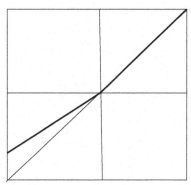

Fig. 4.3  Holonomy of the Reeb foliation.

define the map $\phi : S^1 \times I \to S = \phi(S^1 \times I)$ by

$$\phi(\theta, t) = X_f(c(\theta)).$$

It is clear that $\phi$ is an immersion of class $C^r, r \geq 2$. Then $w^* = \phi^*(w)$ is a well defined 1-form in $S^1 \times I$. Because $dw^* = d\phi(w^*) = \phi^*(dw) = \phi^*(0) = 0$ we have that $w^*$ is closed. Hence $w^*$ defines a foliation $\mathcal{F}^*$ in $S^1 \times I$. Note that $\mathcal{F}^*$ is conjugated to $\mathcal{F} \cap S$. It follows that the curves $c^* = S^1 \times 0$ and $c$ have the same holonomy. Let us calculate the holonomy of $c^*$. Fix $\theta^*, 0) \in c^*$ and $\Sigma^* = \theta^* \times$. Clearly $\Sigma^*$ is a transverse of $\mathcal{F}^*$. Let $f^* : \mathrm{Dom}(f^*) \subset \Sigma^* \to \Sigma^*$ be the holonomy of $c^*$, $p \in \mathrm{Dom}(f^*)$ and $q = f^*(p)$. Let $\alpha$ be the arc in $\Sigma^*$ joining $p$ and $q$.

Let $l$ be the arc in $\mathcal{F}^*$ joining $p, q$. Let $R$ be the closed region bounded by the curves $c^*$, $l$ and $\alpha^*$. Because

$$0 = \int_R dw^* = \int_{\partial R} w^* = \int_l w^* + \int_\alpha w^* = 0 + 0 + \int_\alpha w^*$$

one has

$$\int_\alpha w^* = 0.$$

This equality implies that $\alpha$ is trivial and so $p = q = f^*(q)$. We conclude that $c^*$ has trivial holonomy. Hence $c$ has trivial holonomy and the proof follows.

**Example 4.5 (Holonomy of suspended foliations).** The suspension of a group action is introduced in Section 1.2.6. Regarding the holonomy

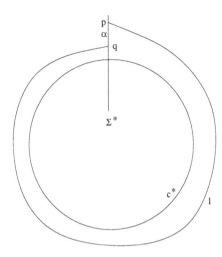

Fig. 4.4

of the resulting foliation we have:

**Theorem 4.1.** *Let $B \times_Q F$ be the suspension of a representation*

$$Q : \pi_1(B) \to \mathrm{Diff}^r(F)$$

*and $\mathcal{F}_Q, \mathcal{F}_Q'$ the corresponding foliations. Then,*

*(1) $\mathcal{F}_Q'$ is a foliation without holonomy.*
*(2) $\mathrm{Hol}(L) \approx Q(\pi_1(B))$, $\forall$ leaf $L$ of $\mathcal{F}_Q$.*

**Proof.** Let $B, F$ be smooth manifolds and $M = B \times_Q F$ be the suspension of a representation $Q : \pi_1(B) \to \mathrm{Diff}(F)$. Recall that $M$ is equipped with two foliations $\mathcal{F}_Q$ and $\mathcal{F}_Q'$ which are the projection over $B \times_Q F$ of the trivial foliations $\{\widetilde{B} \times *\}$ and $\{* \times F\}$ on $\widetilde{B} \times F$ respectively. Because the foliation $\mathcal{F}_Q'$ is induced by a fibration (with fiber $F$) we can see that $\mathcal{F}_Q'$ has no holonomy. So, it is enough to study the holonomy of $\mathcal{F}_Q$. For this we fix a leaf $L$ of $\mathcal{F}_Q$ and choose $x_0 \in L$. Fix $(\widetilde{b}_0, f_0) \in \pi^{-1}(x_0)$, where $\pi : \widetilde{B} \times F \to B \times_Q F$ is the natural projection. It follows from the definition of $\mathcal{F}_Q$ that $L = \pi(\widetilde{B} \times f_0)$. Let $c_0$ be a closed curve containing $x_0$. We want to calculate the holonomy $h_0 : \mathrm{Dom}(h_0) \subset \Sigma_0 \to \Sigma_0$ of $c_0$ in $L$, where $\Sigma_0$ is a suitable transverse containing $x_0$. For this purpose we choose $\Sigma_0 = \pi(\widetilde{b}_0 \times F)$.

**Remark 4.1.**

$\pi\big|_{(\widetilde{B}\times f_0)} : \widetilde{B} \times f_0 \to L$ is a cover map.

In fact observe that

$$\pi(\widetilde{b}, f_0) = \pi(\widetilde{d}, f_0) \Rightarrow \{(g\widetilde{b}, Q(g)f_0) : g \in \pi_1(B)\}$$

$$= \{(g\widetilde{d}, Q(g)f_0) : g \in \pi_1(B)\}$$

$$\Rightarrow \{g\widetilde{b} : g \in \pi_1(B)\} = \{g\widetilde{d} : g \in \pi_1(B)\}.$$

Hence if $P : \widetilde{B} \to B$ is the universal cover of $B$ then $P(\widetilde{b}) = P(\widetilde{d})$ proving the result.

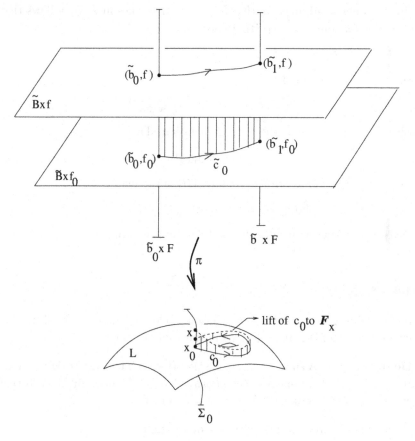

Fig. 4.5

By the previous remark we can consider the lift $\widetilde{c}_0$ of $c_0$ in $\widetilde{B} \times f_0$ with $\widetilde{c}_0(0) = (\widetilde{b}_0, f_0)$. We can write $\widetilde{c}_0(t) = (\widetilde{\gamma}(t), f_0)$. Define $(\widetilde{b}_1, f_0) = \widetilde{c}_0(1)$. Fix $x \in \Sigma_0$ and

$$(\widetilde{b}_0, f) \in \pi^{-1}(x).$$

The curve

$$c(t) = \pi(\widetilde{\gamma}(t), f)$$

is the lift of $c_0$ to the leaf $\mathcal{F}_x$ of $\mathcal{F}$ containing $x$. Hence

$$h_0(x) = \pi(\widetilde{b}_1, f).$$

On the other hand, observe that $\widetilde{b}_1 = \widetilde{\gamma}(1)$ and $\widetilde{\gamma}$ lies in $\widetilde{B}$. It follows that $\widetilde{b}_1 = g_0 \widetilde{b}_0$ for some $g_0 \in \pi_1(B)$. Hence

$$h_0(x) = \pi(\widetilde{b}_1, f) = \pi(g_0 \widetilde{b}_0, f) = \pi(\widetilde{b}_0, Q(g_0^{-1})f).$$

Since $x = \pi(\widetilde{b}_0, f)$ one has

$$h_0 \circ (\pi \circ i) = (\pi \circ i) \circ Q(g_0^{-1}),$$

where $i : F \to \widetilde{b}_0 \times F$ is the natural inclusion. Then,

$$h_0 = \Phi(Q(g)),$$

where $g = g_0^{-1}$ and $\Phi : \mathrm{Hol}(L) \to Q(\pi_1(B))$ is the map defined by

$$\Phi(Q(g)) = (\pi \circ i) \circ Q(g_0^{-1}) \circ (\pi \circ i)^{-1}.$$

One can prove without difficulty that $\Phi$ is an isomorphism. This proves the result. □

## 4.2 Stability

In this section we consider $\mathcal{F}$ a foliation of class $C^1$ of a manifold $M$. We introduce one of the main notions in the study of foliations.

**Definition 4.2.** A subset $B \subset M$ is *stable* (for $\mathcal{F}$) if for every neighborhood $W$ of $B$ in $M$ there exists a neighborhood $W' \subset W$ of $B$ in $M$ such that every leaf of $\mathcal{F}$ intersecting $W'$ is contained in $W$.

A classical problem in the theory of foliations is:

**Problem 4.1.** *Find necessary and sufficient conditions for $B \subset M$ to be stable.*

**Exercise 4.2.1.** Prove that if $M$ is compact then all stable sets of $\mathcal{F}$ are $\mathcal{F}$-invariant.

**Exercise 4.2.2.** Prove that $W'$ in the definition of stable set can be assumed to be invariant.

**Example 4.6.** Let $w$ be the 1-form in $M = \mathbb{R}^2 \setminus \{(0,0)\}$ defined by $w = x\,dx + y\,dy$. Clearly $w = df$ where $f(x,y) = \frac{x^2 + y^2}{2}$ and so $w$ is exact. Hence $w$ is tangent to a foliation $\mathcal{F}$ whose leaves are concentric circles around $(0,0)$. Clearly every leaf of $\mathcal{F}$ is stable, has infinite fundamental group and finite holonomy group.

**Example 4.7.** The compact leaf of the Reeb foliation in $S^3$ is not stable and has infinite holonomy group.

The above example shows the relation between stability and the finiteness of the holonomy group. This relation is the content of the following result, a first step towards Reeb local stability theorem (Theorem 4.3), which is landmark in the theory of foliations.

**Theorem 4.2.** *A compact leaf with finite holonomy group is stable.*

**Proof.** Let $F$ be a compact leaf of a $C^1$ foliation $\mathcal{F}$ with finite holonomy group $\mathrm{Hol}(F)$. We fix $x_0 \in F$ a base point and a transverse $\Sigma_0$ of $\mathcal{F}$ with $\Sigma_0 \cap F = \{x_0\}$. By assumption there are closed curves $\beta_1, \cdots, \beta_k$ containing $x_0$ such that

$$\mathrm{Hol}(F) = \{[f_{\beta_1}], \cdots, [f_{\beta_k}]\},$$

where $f_{\beta_i} : \mathrm{Dom}(f_{\beta_i}) \subset \Sigma_0 \to \Sigma_0$ is the holonomy of $\beta_i$ and $[\cdot]$ denotes the class in the space of germs of $\Sigma$ at $x_0$.

Fix a finite cover $\mathcal{U} = \{U_1, \cdots, U_r\}$ of $M$ by charts of $\mathcal{F}$ such that

(1) $U_i \cap F$ is a plaque $\alpha_i$, $\forall i$;
(2) $U_i \cap U_j \neq \emptyset \Leftrightarrow \alpha_i \cap \alpha_j \neq \emptyset$, $\forall i, j$;
(3) each plaque of $U_i$ intersects at most one plaque of $U_j$, $\forall i, j$.

The following notation will be useful:

- $\Sigma_i =$ space of plaques of $U_i$.
- $\pi_i : U_i \to \Sigma_i$ the plaque projection.
- $x_i = \pi_i(\alpha_i)$. We can suppose that $x_i$ is a point of $\alpha_i$.

In the case $U_i \cap U_j \neq \emptyset$ ($\Leftrightarrow \alpha_i \cap \alpha_j \neq \emptyset$) we let $\bar{c}_{i,j} \subset \alpha_i \cup \alpha_j$ be a curve joining $x_i$ and $x_j$. Also in this case we let $\gamma_{i,j} : \mathrm{Dom}(\gamma_{i,j}) \subset \Sigma_i \to \Sigma_j$ be the holonomy along the plaques. For all $i = 2, \cdots, r$ we let $c_i \subset F$ be a curve joining $x_1$ with $x_i$. The curve $c_i$ induces a holonomy $h_i : \mathrm{Dom}(h_i) \subset \Sigma_1 \to \Sigma_i$.

To prove the stability of the leaf $F$ we fix a neighborhood $W$ of $F$. Without loss of generality we can assume: $\cup_1^r U_i \subset W$, $\Sigma_1 \subset \Sigma_0$, $x_1 = x_0$ and $D := \cap_1^k \mathrm{Dom}(f_{\beta_i})$ to be a neighborhood of $x_0$ in $\Sigma_1$. The closed curves of the form

$$c_{j,l} = c_l \cup \bar{c}_{l,j} \cup c_j.$$

induce a holonomy map

$$h_{j,l} : \mathrm{Dom}(h_{j,l}) \subset \Sigma_1 \to \Sigma_1.$$

Since $\mathrm{Hol}(F)$ is the union of $[f_{\beta_i}]$'s we have that $h_{l,j} = f_{\beta_p}$ in a neighborhood $D_{j,l} \subset \Sigma_1$ of $x_1(= x_0$ the base point). Define

$$D^* = \cap_{j,l} D_{j,l}.$$

Then $D^*$ is an open neighborhood of $x_0$ in $\Sigma_1$. Of course $D^* \subset D$ and $h_{j,l} = f_{\beta_p}$ in $D^*$. Let $D' \subset D^*$ be a neighborhood of $x_1$ such that $y' \in D' \Rightarrow f_{\beta_i}(y) \in D, \forall i$.

For every $y \in D'$ we define

$$\mathcal{C}_y^* = \{\pi_j^{-1}(h_j(f_{\beta_i}(y))) : 1 \leq i \leq k, 1 \leq j \leq r\},$$

where $h_1 =$ Identity by definition.

**Claim 4.1.** *If $P$ is a plaque of $U_l$ (for some $l = 1, \cdots, r$) and $P \cap L \neq \emptyset$ for some $L \in \mathcal{C}_y^*$, then $P \in \mathcal{C}_y^*$.*

**Proof of Claim 4.1:** By hypothesis there is $L = \pi_j^{-1}(h_j(f_{\beta_i}(y)))$ such that $P \cap L \neq \emptyset$. Because $L \subset U_j$ (as it is a plaque of $U_j$) and $P \subset U_l$ we have that $U_j \cap U_l \neq \emptyset$. This implies that $c_{j,l}$ is well defined (recall that this is a curve joining $x_i$ with $x_j$ in $\alpha_i \cap \alpha_j$). By the definition of the holonomy $\gamma_{j,l}$ we have

$$\gamma_{j,l}(h_j(f_{\beta_i}(y))) = \pi_l(P).$$

Hence

$$h_l^{-1} \circ \gamma_{j,l} \circ h_j(f_{\beta_i}(y)) = h_l^{-1}(\pi_l(P)).$$

But, by definition, $h_l^{-1} \circ \gamma_{j,l} \circ h_j$ is precisely the holonomy $h_{j,l}$ of the curve $c_{j,l}$. Hence

$$h_{j,l}(f_{\beta_i}(y)) = h_l^{-1}(\pi_l(P)).$$

Because $h_{j,l} = f_{\beta_p}$ one has

$$f_{\beta_p} \circ f_{\beta_i}(y) = h_l^{-1}(\pi_l(P)).$$

As $\mathrm{Hol}(F)$ is a group one has $f_{\beta_p} \circ f_{\beta_i} = f_{\beta_{i'}}$ for some $i'$. Hence

$$f_{\beta,i'}(y) = h_l^1(\pi_l(P)) \Rightarrow \pi_l^{-1}(h_l(f_{\beta_{i'}}(y))) = P,$$

proving $P \in C_y^*$ as desired. This proves Claim 4.1. $\qquad\square$

Now, we let

$$L_y^* = \cup\{L : L \in C_y^*\}.$$

We have the following properties: $L_y^* \subset \mathcal{F}_y$(=the leaf of $\mathcal{F}$ containing $y$); $L_y^*$ is open in $\mathcal{F}_y$ (since it is union of plaques of $\mathcal{F}_y$); $L_y^*$ is close in $M$ (this is Claim 4.1); $L_y^* = \mathcal{F}_y$ (because $\mathcal{F}_y$ is connected); $L_y^* \subset W$ (because the plaques forming $C_y^*$ are contained in $\bigcup_1^r U_i \subset W$). The last two properties above imply that $\mathcal{F}_y \subset W$, $\forall y \in D'$. Defining $W'$ as the set of leaves intersecting $D'$ we have that $W' \subset W$ is a neighborhood of $F$ such that every leaf of $\mathcal{F}$ intersecting $W'$ is contained in $W$. Since $W$ is arbitrary the result follows. $\qquad\square$

Exercise 4.2.3. (Prove or give a counterexample) A compact invariant set whose leaves have finite holonomy group is stable.

## 4.3 Reeb stability theorems

The statement of Theorem 4.2 can be improved as follows.

**Theorem 4.3 (Reeb Local Stability Theorem).** *Let $F$ be a compact leaf with finite holonomy group of a $C^r$ foliation $\mathcal{F}$ in a manifold $M$. Then for each neighborhood $W$ of $F$ there is a $C^r$ $\mathcal{F}$-invariant tubular neighborhood $\pi : W' \subset W \to F$ of $F$ with the following properties:*

*(1) Every leaf $F' \subset W'$ is compact with finite holonomy group.*
*(2) If $F' \subset W'$ is a leaf then the restriction $\pi/F' : F' \to F$ is a finite cover map.*
*(3) If $x \in F$ then $\pi^{-1}(x)$ is a transverse of $\mathcal{F}$.*

**Proof.** Let $W$ be a fixed neighborhood of $F$. We can assume that $W$ is the domain of a $C^r$ tubular neighborhood $\pi_0 : W \to F$. Because $F$ is compact we can further assume that the fiber $\pi_0^{-1}(x)$ is transverse to $\mathcal{F}$, $\forall x \in F$. Let $W' \subset W$ be given by Theorem 4.2. It follows from the proof of Theorem 4.2 that all leaves $F'$ in $W'$ are compact (all of them have the form $F' = L_y^*$ for some $y \in D'$ and $L_y^*$ is compact). Define $\pi = \pi_0/W'$. Then $\pi : W' \to F$ is a tubular neighborhood which is invariant and satisfies (3). By shrinking $W'$ if necessary we can assume that $F'$ is transverse to the fiber $\pi^{-1}(x)$, $\forall x \in F$. Since all leaf $F' \subset W'$ is compact we have that $F'$ intersect each fiber finitely many times. The same argument shows that every leaf $F' \subset W'$ has finite holonomy group. This proves (1) and (2). The theorem is proved. $\qquad \square$

Next we state two useful lemmas.

**Lemma 4.1.** *Let* $\mathrm{Hom}(\mathbb{R}, 0)$ *be the group of germs of homeomorphisms in* $\mathbb{R}$ *fixing* $0 \in \mathbb{R}$. *If* $G$ *is a finite subgroup of* $\mathrm{Hom}(\mathbb{R}, 0)$ *then* $G$ *has at most two elements. If all the elements of* $G$ *are represented by orientation-preserving maps, then* $G = \{[Id]\}$.

**Proof.** Suppose that there is $[f] \in G - \{[Id]\}$ represented by a local orientation-preserving homeomorphism $f$ fixing 0. On one hand, there are $n_0 \in \mathbb{N}$ and a neighborhood $U \subset \mathbb{R}$ of 0 such that $f^{n_0}(x) = x$ for all $x \in U$ because $[f]$ has finite order in $G$ (as $G$ is finite). On the other hand, there is $x_0 \in U$ such that $f(x_0) \neq x_0$ because $[f] \neq [Id]$. We can suppose that $0 < x_0$ and that $[0, x_0] \subset U$ without loss of generality. Because $f$ is orientation-preserving one has $0 < f^n(x_0) < f^{n-1}(x_0) < \cdots < f^{(}x_0) < x_0$ for all $n \in \mathbb{N}$ Clearly $f^n(x_0) \in [0, x_0]$ for all $n$ as $[0, x_0] \subset U$. The last applied to $n = n_0$ yields $f^{n_0}(x_0) = x_0$ and so $x_0 < x_0$, a contradiction. This contradiction shows that $[f] = [Id]$ for all element $[f] \in G$ represented by a local orientation-preserving homeomorphism $f$ fixing 0. Let $[g], [g'] \in G$ be represented by orientation-reversing local homeomorphisms fixing 0. Hence $[g] \cdot [g']^{-1}$ is represented by $g \circ (g')^{-1}$ which is orientation-preserving. It follows that $[g] = [g']$ and so there is only one element of $G$ represented by an orientation-reversing map. This proves that $G$ has at most two elements and the proof follows. $\qquad \square$

**Lemma 4.2.** *Let* $F$ *be a compact leaf of a codimension one foliation* $\mathcal{F}$ *defined on a manifold* $M$. *Let* $F_n$ *be a sequence of compact leaf of* $\mathcal{F}$ *accumulating to a point in* $F$. *Then* $\forall$ *neighborhood* $W \subset M$ *one has* $F_n \subset W$ *for all* $n$ *large.*

**Proof.** Let $U_1, \cdots, U_k \subset W$ be cover of $F$ with charts of $\mathcal{F}$ such that $U_i \cap F$ is a single plaque $\alpha_i$ of $U_i$, $\forall i$. For each $i$ we denote by $\Sigma_i$, the space of plaques of $U_i$, and by $\pi_i : U_i \to \Sigma_i$ the projection along the plaques. Because $F_n$ accumulates a point of $F$ we can assume that $F_n \cap U_1$ contains a plaque arbitrarily close to $\alpha_1$. From this we can assume that $F_n \cap U_i \neq \emptyset$ for all $n, i$. Clearly $F_n \cap U_i$ contains a finite number of plaques as $F_n$ is compact. Let $P^{n,i}$ and $p^{n,i}$ be the maximum and the minimum of such plaques with respect to the natural order of $\Sigma_i$ (=interval). Define $R_n = \cup_{i=1}^{k}(P^{n,i} \cup p^{n,i})$. Clearly $R_n \subset F_n$ is open in $F_n$ (as it is union of plaques). Let us prove that $R_n$ is closed in $F_n$. In fact, fix $n$ and choose a sequence $x_j \in R_n$ converging to $x \in F$. We can assume that all the $x_j$'s are in a single plaque $P^{n,i_0}$ of $R_n$. Clearly $x \in U_r$ for some $1 \leq r \leq k$ by the definition of $R_n$. Hence the plaque $\alpha_r(x) \subset U_r$ containing $x$ is well defined. Clearly $P^{n,i_0} \cap U_r \neq \emptyset$ and so $P^{n,i_0} \cap \alpha_r(x) \neq \emptyset$. Thus $\alpha_r(x)$ is a plaque of $F_n \cap U_r$. Since $P^{n,i_0}$ is the maximum of the plaques of $F_n \cap U_{i_0}$ one has that $\alpha_r(x)$ is the maximum of the plaques of $F_n \cap U_r$. In other words $\alpha_r(x) = P^{n,r}$ proving $x \in R_n$. We conclude that $R_n$ is closed in $F_n$. Since $F_n$ is connected we conclude that $R_n = F_n$. Since $R_n \subset \cup_i U_i \subset W$ we conclude that $F_n \subset W$. The lemma is proved. $\square$

**Remark 4.2.** The conclusion of the lemma above is false for foliations of codimension $> 1$.

For the case of codimension one foliations, the conclusion of the local stability can be reinforced, by assuming that the compact leaf has finite fundamental group. In this case, if the manifold is also compact, then all leaves of the foliation are compact.

**Theorem 4.4 (Reeb global stability theorem).** *Let $\mathcal{F}$ be a $C^1$ codimension one foliation of a closed manifold $M$. If $\mathcal{F}$ contains a compact leaf $F$ with finite fundamental group then all the leaves of $\mathcal{F}$ are compact with finite fundamental group. If $\mathcal{F}$ is transversely orientable then every leaf of $\mathcal{F}$ is diffeomorphic to $F$; $M$ is the total space of a fibration $f \colon M \to S^1$ over $S^1$ with fiber $F$; and $\mathcal{F}$ is the fiber foliation $\{f^{-1}(\theta) : \theta \in S^1\}$.*

**Proof.** Denote by $\mathcal{F}_x$ the leaf of $\mathcal{F}$ containing $x \in M$ and define

$$\hat{M} = \{x \in M : \mathcal{F}_x \text{ is compact with } \pi_1 \text{ finite}\}.$$

Note that by hypothesis $W \neq \emptyset$. The Reeb local stability theorem implies that $\hat{M}$ is open. We can assume that $\hat{M}$ is connected (otherwise we replace it by a connected component). Let us prove that $\hat{M}$ is closed. For this it suffices to prove that $\partial\hat{M} = \emptyset$. Suppose by contradiction that there exists $x_0 \in \partial\hat{M}$. Let $U$ be a chart of $\mathcal{F}$ containing $x_0$, $\Sigma$ the space of plaques of $U$ and $\pi : U \to \Sigma$ be the projection along the plaques. Note that $\Sigma$ is an interval (as $\mathcal{F}$ has codimension one) and $\hat{M} \cap U$ is union of plaques of $\mathcal{F}$. It follows that $\pi(\hat{M} \cap V) \subset \Sigma$ is a countable family of open intervals. Let $J$ be one of these intervals and $\mathcal{F}(J)$ be the union of the leaves of $\mathcal{F}$ intersecting $J$. Because $J$ is open and contained in $\hat{M}$ we have that $\mathcal{F}(J)$ is open in $\hat{M}$. We claim that $\mathcal{F}(J)$ is closed in $\hat{M}$. Indeed, consider a sequence $\mathcal{F}(J) \ni x_n \to x \in \hat{M}$. Assume by contradiction that $x \notin \mathcal{F}(J)$. Since $x \in \hat{M}$ the Reeb local stability theorem implies that there is a neighborhood $R$ of $\mathcal{F}_x$ such that every leaf intersecting $R$ is compact with $\pi_1$ finite and contained in $R$. On one hand we can choose $R$ such that $R \cap J = \emptyset$. On the other hand we observe that $\mathcal{F}_{x_n} \cap R \neq \emptyset$ for large $n$ because $x_n \to x$. Hence $\mathcal{F}_{x_n} \subset R$ and also $\mathcal{F}_{x_n} \cap J \neq \emptyset$ by the definition of $\mathcal{F}(J)$. Hence $R \cap J \neq \emptyset$ a contradiction. This contradiction shows that $x \in \mathcal{F}(J)$, *i.e.*, $\mathcal{F}(J)$ is closed. The claim follows. By connectedness we conclude that $\mathcal{F}(J) = \hat{M}$. It follows that every leaf of $\mathcal{F}$ in $\hat{M}$ intersect $J$. Since every leaf of $\mathcal{F}$ in $\hat{M}$ is compact we conclude that $\pi(\hat{M} \cap U)$ is a finite union of open intervals in $\Sigma$. We conclude that $x$ is a boundary point of one of these intervals. Now we claim that the leaf $\mathcal{F}_x$ is closed. Otherwise the above argument would imply that there exist a foliated chart $U$ such that $\hat{M} \cap U$ contains infinitely many connected components, a contradiction. Because $M$ is compact by assumption we conclude that $\mathcal{F}_x$ is compact. Hence there is a tubular neighborhood $P : W \to F$ of $F$ whose fibers $P^{-1}(y)$, $y \in F$ are transverse to $\mathcal{F}$. Because $x \in \partial\hat{M}$ there is a sequence of compact leaves $F_n$ with finite $\pi_1$ accumulating on $x$. By Lemma 4.2 we can assume that $F_n \subset W$. The restriction $P/F_n : F_n \to F$ is a finite cover of $F$. Because $F_n$ has finite fundamental group and $P/F_n : F_n \to F$ is a finite cover we conclude that $F$ has finite $\pi_1$. We conclude that $x \in \hat{M}$ contradicting $x \in \partial\hat{M}$. This contradiction proves that $\partial\hat{M} = \emptyset$ proving that $\hat{M}$ is closed. By connectedness reasons we conclude that $M = \hat{M}$. Hence all leaves of $\mathcal{F}$ are compact with finite $\pi_1$.

Now suppose that $\mathcal{F}$ is transversely orientable. We already know that each leaf $L$ of $\mathcal{F}$ is compact with finite $\pi_1$. Being $\mathcal{F}$ transversely orientable we have that the holonomy group of $L$ is represented by an orientation-preserving homomorphism. In other words the subgroup $G = \text{Hol}(L)$ of $\text{Diff}^0(\mathbb{R}, 0)$ is formed by orientation-preserving maps. Then Lemma 4.1 implies that $\mathcal{F}$ is a foliation without holonomy. It follows from the proof of the Reeb local stability theorem that $\mathcal{F}$ is *locally a product foliation*, *i.e.*, each leaf $L$ of $\mathcal{F}$ is equipped with an invariant product neighborhood $L \times I$ such that the leaves of $\mathcal{F}$ in this neighborhood have the form $L \times *$,

$* \in I$, with $L \times 0$ corresponding to $L$. On the other hand, $M$ is compact by hypothesis. Hence there is a closed curve $c$ transverse to $\mathcal{F}$ (to find $c$ we simply use the non-wandering set of the transverse vector field associated to $\mathcal{F}$). Moreover $c$ can be chosen to intersect some leaf $L_0$ of $\mathcal{F}$ in a single point. Observe that $c$ intersect all leaves of $\mathcal{F}$ in a single point. In fact, consider the set $c' = \{x \in c : \mathcal{F}_x \cap c = \{x\}\}$. The set $c'$ is not empty by the existence of $L_0$. The fact that $\mathcal{F}$ is a locally product foliation implies that $l$ is open and closed in $c$, therefore $c' = c$ and so $c$ intersect each leaf of $\mathcal{F}$ in one point at most. Now let $\mathcal{F}(c)$ be the set of points $x \in M$ such that $\mathcal{F}_x \cap c \neq \emptyset$. It is clear that $\mathcal{F}(c)$ is open. One can prove that $\mathcal{F}(c)$ is closed by using the Reeb local stability theorem as before. Hence $\mathcal{F}(c) = M$ proving that all leaves of $\mathcal{F}$ intersect $c$. To define the desired fibration $f \colon M \to S^1$ we simply define $f(x)$ to be the intersection point of $\mathcal{F}_x$ with $c$. The theorem is proved. $\qquad \square$

**Exercise 4.3.1.** Prove that there is no codimension one foliation in the closed 3-ball $B^3$ having $\partial B^3 = S^3$ as a leaf.

**Exercise 4.3.2.** (Prove or give a counterexample) Codimension one transitive foliations have no compact leaves.

**Exercise 4.3.3.** Show that if $G$ is a simply-connected Lie group and $M$ is a compact manifold of dimension $\dim M = 1 + \dim G$ then for $\dim G \geq 2$ there is no locally free action of $G$ in $M$.

**Exercise 4.3.4.** Let $\mathcal{F}$ be a transversely orientable codimension one foliation on a closed orientable 3-manifold $M$. If there is a leaf $F$ of $\mathcal{F}$, whose universal cover is not the real plane $\mathbb{R}^2$ then $M = S^2 \times S^1$ and $\mathcal{F}$ is the product foliation $S^2 \times *$. What about the case $M$ is not orientable?

**Exercise 4.3.5.** Let $\omega$ be a $C^2$ integrable 1-form in a neighborhood of the origin $0 \in \mathbb{R}^n$. We assume that the origin is a singularity of center type for $\omega$ so that, up to a linear change of coordinates we have $\omega = d(\frac{1}{2} \sum_{j=1}^{n} x_j^2) + (\ldots)$ where $(\ldots)$ means higher order terms. A classical result due to Reeb states that for $n \geq 3$ there is a neighborhood of the origin where all the leaves of $\mathcal{F}_\omega : \omega = 0$ are diffeomorphic to the $(n-1)$-sphere. This is proved as follows:

(i) Consider the cylindrical blow-up of the origin given by the map $\sigma \colon \mathbb{R} \times S^{n-1} \to \mathbb{R}^n$, $\sigma(t,x) = t.x$. Show that $\{0\} \times S^{n-1}$ is a leaf of the lifted foliation $\mathcal{F}^* = \sigma(\mathcal{F}_\omega)$ (hint: show that the 1-form $\Omega^* = \frac{1}{t}\sigma^*(\omega)$ defined in

$(\mathbb{R} - \{0\}) \times S^{n-1}$ extends to $\mathbb{R} \times S^{n-1}$ as $\Omega^* = dt$ for $t = 0$ in class $C^1$. Also show that $\{0\} \times S^{n-1}$ is a leaf of $\sigma^*(\omega)$ and so of $\Omega^*$.

(ii) For $n \geq 3$ use the Reeb local stability theorem to conclude.

**Exercise 4.3.6.** Look (possibly in the literature) for a demonstration of the following analytic version (also due to Reeb) of the above exercise: If $\omega$ is a real analytic integrable 1-form in a neighborhood of the origin $0 \in \mathbb{R}^n$ and $n \geq 3$. Suppose that the linear part of $\omega$ is non-degenerate and $\omega = df + (...)$ for some quadratic analytic function $f$. Then there is a neighborhood of the origin where $\mathcal{F}_\omega : \omega = 0$ is analytically conjugate to the linear foliation $df = 0$.

## 4.4    Thurston stability theorem

In this section we prove the following theorem known as the *Thurston stability theorem* [Thurston (1974)].

**Theorem 4.5.** *Let $\mathcal{F}$ be a $C^1$ transversely orientable codimension one foliation of a compact manifold $M$ tangent to the boundary of $M$ if nonempty. If $\mathcal{F}$ has a compact leaf $L$ with trivial first cohomology group (over the reals), then every leaf of $\mathcal{F}$ is homeomorphic to $L$ and, furthermore, $M$ is homeomorphic either to $L \times [0,1]$ with $\mathcal{F}$ being the product foliation, or, to the total space of a fibration over $S^1$ having the leaves of $\mathcal{F}$ as fibers.*

The proof we give here is the one by Schachermayer [Schachermayer (1978)] which in turns is a simplification of Reeb and Schweitzer [Reeb and Schweitzer (1978)]. It is based on the elementary lemma below.

**Lemma 4.3.** *Let $\epsilon > 0$ and $f, g, h : (-\epsilon, \epsilon) \to \mathbb{R}$ be $C^1$ orientation-preserving embeddings such that $f(0) = g(0) = h(0) = 0$ and $f'(0) = g'(0) = h'(0) = 1$. If $(x_n)_{n \in \mathbb{N}^+} \subseteq (-\epsilon, \epsilon)$ is a sequence such that the limits*

$$\lim_{n \to \infty} \frac{f(x_n) - x_n}{h(x_n) - x_n} = A \quad and \quad \lim_{n \to \infty} \frac{g(x_n) - x_n}{h(x_n) - x_n} = B$$

*exist, then*

$$\lim_{n \to \infty} \frac{f(g(x_n)) - x_n}{h(x_n) - x_n} = A + B \quad and \quad \lim_{n \to \infty} \frac{f^{-1}(x_n) - x_n}{h(x_n) - x_n} = -A.$$

**Proof.** Applying the Mean Value theorem we get

$$f(g(x_n)) - x_n = f(x_n) - x_n + f(g(x_n)) - f(x_n) = f(x_n) - x_n + f'(\xi_n) \cdot (g(x_n) - x_n)$$

for some $\xi_n$ in between $g(x_n)$ and $x_n$. Dividing by $h(x_n) - x_n$ and letting $n \to \infty$ we get the first limit.

Again the Mean Value theorem implies

$$f^{-1}(x_n) - x_n = -(f^{-1}(f(x_n)) - f^{-1}(x_n)) = -(f^{-1})'(\rho_n) \cdot (f(x_n) - x_n)$$

for some $\rho_n$ in between $f(x_n)$ and $x_n$. Dividing by $h(x_n) - x_n$ and letting $n \to \infty$ we get the second limit. $\qquad \square$

**Proof of Theorem 4.5.** Suppose by contradiction that the conclusion of the theorem does not hold. Then, by the proof of the Reeb global stability theorem, the holonomy group $\mathrm{Hol}(L)$ of $L$ is nontrivial, *i.e.*, $\mathrm{Hol}(L) \neq \{Id\}$. Put $\mathcal{H}(L) = \mathrm{Hol}(L)$. Let $\hat{h} : \pi_1(L) \to \mathcal{H}(L)$ be the map assigning to a loop $s$ in $L$ its holonomy map $\hat{h}_s$. Since $\mathcal{F}$ is $C^1$, $\hat{h}_s$ is differentiable and we denote by $\hat{h}'_s(0)$ the corresponding derivative at 0. This yields the map $\hat{h} : \pi_1(L) \to \mathbb{R}^+$ assigning to each $s \in \pi_1(L)$ the number $\hat{h}'_s(0)$ which is positive since $\mathcal{F}$ is transversely orientable. The composition $\log \circ \hat{h} : \pi(L) \to \mathbb{R}$ is clearly a group homomorphism. Since $H^1(L, \mathbb{R}) = \mathrm{Hom}(\pi_1(L), \mathbb{R})$ where the later is the group of homomorphism from $\pi_1(L)$ to $\mathbb{R}$, and $H^1(L, \mathbb{R}) = 0$ by hypothesis, we obtain that the homomorphism $\log \circ \hat{h}$ is trivial. Consequently, $\hat{h}'_s(0) = 1$ for every $s \in \pi_1(L)$.

Let $\alpha, \beta, \gamma, \cdots, \lambda : (-\epsilon, \epsilon) \to \mathbb{R}$ be $C^1$ orientation-preserving embeddings fixing 0 whose germs $\hat{\alpha}, \hat{\beta}, \hat{\gamma}, \cdots \hat{\lambda}$ generates $\mathcal{H}(L)$. Because $\mathcal{H}(L) \neq 0$, there is a sequence $(x_n)_{n \in \mathbb{N}^+} \subset (0, \epsilon)$ converging monotonically to 0 such that

$$(\tilde{\alpha}(x_n), \tilde{\beta}(x_n), \tilde{\gamma}(x_n), \cdots, \tilde{\lambda}(x_n)) \neq (0, 0, 0, \cdots, 0),$$

where $\tilde{T}(x) = T(x) - x$ for $T \in \{\alpha, \beta, \gamma, \cdots, \lambda\}$ and $x \in (-\epsilon, \epsilon)$. Define

$$\mu_n = \max\{|\alpha(x_n)|, |\beta(x_n)|, |\gamma(x_n)|, \cdots, |\lambda(x_n)|\}$$

and

$$N_T = \{n \geq 1 : |\tilde{T}(x_n)| = \mu_n\} \quad \text{for} \quad T \in \{\alpha, \beta, \gamma, \cdots, \lambda\}.$$

It follows that $\mathbb{N}^+ = N_\alpha \cup N_\beta \cup N_\gamma \cup \cdots \cup N_\lambda$. Reordering the generators if necessary we can assume $0 \neq |\tilde{\alpha}(x_n)| = \mu_n$ for all $n \in \mathbb{N}^+$. Since

$$\left| \frac{\tilde{\beta}(x_n)}{\tilde{\alpha}(x_n)} \right| \leq 1, \quad \cdots, \left| \frac{\tilde{\lambda}(x_n)}{\tilde{\alpha}(x_n)} \right| \leq 1$$

we can assume (again by passing to a subsequence if necessary) that the limits

$$a = \lim_{n \to \infty} \frac{\tilde{\alpha}(x_n)}{\tilde{\alpha}(x_n)}, \quad b = \lim_{n \to \infty} \frac{\tilde{\beta}(x_n)}{\tilde{\alpha}(x_n)}, \cdots, l = \lim_{n \to \infty} \frac{\tilde{\lambda}(x_n)}{\tilde{\alpha}(x_n)}$$

exist. These limits are not all zero since $a = 1$.

On the other hand, since $\hat{h}'_s(0) = 1$ for every $s \in \pi_1(L)$, we have $T'(0) = 1$ for all $T \in \{\alpha, \beta, \gamma, \cdots, \lambda\}$. Then, by Lemma 4.3, the assignments

$$\hat{\alpha} \mapsto a, \quad \hat{\beta} \mapsto b, \quad \cdots, \hat{\lambda} \mapsto l$$

extends uniquely to an additive homomorphism $\rho : \mathcal{H}(L) \to \mathbb{R}$ which is non-trivial since $a = 1$. Then, the composition $\varphi = \rho \circ \hat{h}$ yields a homomorphism $\varphi : \pi_1(L) \to \mathbb{R}$ which is nontrivial since $\hat{h}$ is onto and $\rho$ is nontrivial. It follows that $H^1(L, \mathbb{R}) = \mathrm{Hom}(\pi_1(L), \mathbb{R}) \neq 0$ a contradiction. This finishes the proof. □

**Exercise 4.4.1.** Let $\mathcal{F}$ be an orientable foliation with a compact leaf $L \in \mathcal{F}$ homologous to zero. Prove that the Euler characteristic of $L$ is zero.

**Exercise 4.4.2.** Let $\mathcal{F}$ be a foliation on $M$ of codimension $q$. A differentiable map $f \colon N \to M$ is *transverse* to $\mathcal{F}$ if it is transverse to each leaf $L \in \mathcal{F}$ as an immersed submanifold in $M$. Show that in this case there is a naturally defined foliation $f^*(\mathcal{F})$ in $N$ of codimension $q$ such that for each leaf $L \in \mathcal{F}$ the inverse image $f^{-1}(L)$ is a union of leaves of $f^*(\mathcal{F})$.

**Exercise 4.4.3.** Show that if $\mathcal{F}$ is a codimension 1 smooth foliation of a manifold $M$ and $L \in \mathcal{F}$ is a compact leaf with $\mathrm{Hom}(\pi_1(L), \mathbb{R}) = 1$ and $H^1(L, \mathbb{R}) = 0$ then $L$ has trivial holonomy.

# Chapter 5

# Haefliger's theorem

## 5.1 Statement

A very nice example of the interplay between Dynamical Systems and Topology in the Theory of foliations is given by Haefliger's theorem. In few words, it says that an analytic foliation of codimension one, cannot exhibit a closed transverse section homotopic to zero in the ambient manifold. The reason is that such a null-homotopic transverse curve will imply the existence of a kind of "limit cycle" for the foliation.

**Definition 5.1.** Let $\mathcal{F}$ be a codimension one foliation on a manifold $M$. A leaf $F$ of $\mathcal{F}$ has *one-sided holonomy* if there are a closed curve $c \subset F$ and point $x_0 \in c$ whose holonomy map $f \colon \mathrm{Dom}(f) \subset \Sigma \to \Sigma$ on a transverse segment $\Sigma$ intersecting $c$ satisfies the following properties:

(1) $f$ is *not* the identity $Id$ in any neighborhood of $x_0$ in $\Sigma$.
(2) $f = \mathrm{Id}$ in one of the two connected components of $\Sigma \setminus \{x_0\}$.

The graph of $f$ above may be as in Figure 4.3.

**Example 5.1.** A leaf with one-sided holonomy cannot be simply connected. The torus fiber of a torus bundle over $S^1$ is a non-simply connected leaf without one-sided holonomy.

**Example 5.2.** The Reeb foliation in $S^3$ is an example of a codimension one $C^\infty$ foliation on a manifold with finite fundamental group with a one-sided holonomy leaf.

**Example 5.3.** Real analytic codimension one foliations cannot have one-sided holonomy leaves.

The main result of this section gives a sufficient condition for the existence of one-sided holonomy leaves.

**Theorem 5.1 (Haefliger's theorem).** *A codimension one $C^2$ foliation with a null-homotopic closed transversal has some leaf with one-sided holonomy.*

**Corollary 5.1.** *Codimension one $C^2$ foliations on compact manifolds with finite fundamental group have one-sided holonomy leaves. In particular, there are no real analytic codimension one foliations on manifolds with finite fundamental group.*

In fact, every codimension one foliation on a compact manifold has a closed transverse. If the fundamental group of the manifold is finite then a suitable power of this curve (as element of the fundamental group) yields a null-homotopic closed transverse. Then Haefliger's theorem applies. The last conclusion of the above corollary applies to the following case:

**Corollary 5.2.** *There is no real analytic codimension one foliation on $S^3$.*

The proof of Haefliger's theorem is divided in three parts according to the forthcoming sections.

## 5.2   Morse theory and foliations

First we recall some classical basic Morse Theory (cf. [Milnor (1963)]). Let $W$ be a compact 2-manifold with boundary $\partial W$ (possibly empty). Let $f \colon W \to \mathbb{R}$ be a $C^r$ map $r \geq 2$. A point $p \in W$ is a *critical point* of $f$ if $f'(p) = 0$. A critical point $p$ is *non-degenerated* if the second derivative $f''(p)$ is a non-degenerated quadratic form, where

$$f''(p) = \left( \frac{\partial^2 (f \circ x^{-1})(0)}{\partial x_i \partial x_j} \right)_{1 \leq i,j \leq 2}$$

for some coordinate system $(x_1, x_2)$ around $p = (0,0)$. We shall use the following lemma due to Morse.

**Lemma 5.1 (Morse Lemma).** *Let $p$ be a non-degenerated critical point of a $C^r$ map $f \colon W \to \mathbb{R}$, $r \geq 2$. Then there is a coordinate system $(x,y)$ around $p = (0,0)$ such that one of the following alternatives hold:*

*(1)* $f(x,y) = f(0,0) + x^2 + y^2$.
*(2)* $f(x,y) = f(0,0) - x^2 - y^2$.

*(3)* $f(x, y) = f(0, 0) + x^2 - y^2.$

The level curves of the three alternatives above are depicted in Figure 5.1.

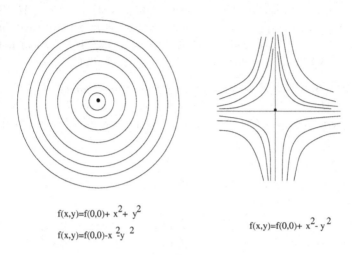

f(x,y)=f(0,0)+ $x^2$+ $y^2$

f(x,y)=f(0,0)-$x^2$-$y^2$

f(x,y)=f(0,0)+ $x^2$- $y^2$

Fig. 5.1

Motivated by the above, we define a *Morse function* as a $C^2$ map $f\colon W \to \mathbb{R}$, such that all of its critical points are non-degenerated. We denote by $C^r(W, \mathbb{R})$ the set of all $C^r$ functions defined on $W$ endowed with the $C^r$ topology (see [Hirsch (1971)]) and by $M^r(W, \mathbb{R}) \subset C^r(W, \mathbb{R})$ the subset of Morse functions.

**Remark 5.1.** Notice that a Morse type critical point is isolated from the set of critical points of a given function. In particular, a Morse function on a compact manifold has only finitely many critical points.

The following is a classical result in Morse Theory and Differential Topology ([Hirsch (1971); Milnor (1963)]).

**Theorem 5.2 (Theorem of Morse, [Hirsch (1971)],[Milnor (1963)]).** *Given a manifold $W$, the set $M^r(W, \mathbb{R})$ is open and dense in $C^r(W, \mathbb{R})$.*

Now we describe the foliated Morse Theory. We let $M$ be a manifold and $\mathcal{F}$ a codimension one foliation of class $C^2$ in $M$.

**Definition 5.2.** A $C^r, r \geq 2$ map $f: W \to M$ is *Morse with respect to $\mathcal{F}$*, if for all $p \in M$ there is a foliated chart $U$ of $\mathcal{F}$ containing $p$ such that if $\pi : U \to \mathbb{R}$ is the projection along the plaques then $\pi \circ f \in M^r(W, \mathbb{R})$. Morse maps with respect to $\mathcal{F}$ are often said to be *in general position with respect to $\mathcal{F}$*. A *critical point of $f$ with respect to $\mathcal{F}$* is a critical point of $\pi \circ f$ for some (and therefore for every) foliated chart $U$.

**Theorem 5.3.** *Let $\mathcal{F}$ be a codimension one foliation of class $C^2$ on a manifold $M$. Let $W$ be a compact 2-manifold and $A : W \to M$ be a $C^r$ map. Then there is a $C^r$ map $f: W \to M$ arbitrarily $C^r$ close to $A$ such that*

*(1) $f$ is Morse with respect to $\mathcal{F}$.*

*(2) If $p, p'$ are different critical points of $f$ with respect to $\mathcal{F}$ then $f(p)$ and $f(p')$ are in different leaves of $\mathcal{F}$.*

**Proof.** Fix an open cover $Q_1, \cdots, Q_k$ of $A(W)$ by foliated charts of $\mathcal{F}$. We can assume without loss of generality that the chart $\phi^i : Q_i \to D^{n-1} \times D^1$ has the form

$$\phi^i = (\phi_1^i, \cdots, \phi_{n-1}^i, \pi_i),$$

where the last coordinate $\pi_i : Q_i \to \mathbb{R}$ denotes the projection along the plaques.

Define $W_i = A^{-1}(U_i)$ for all $i$. Hence $W_1, \cdots, W_k$ is an open cover of $W$. Fix $U_i \subset \overline{U_i} \subset V_i \subset \overline{V_i} \subset W_i$ such that $U_1, \cdots U_k$ is an open cover of $W$. For each $i$ we fix a $C^\infty$ function $\lambda_i : W \to \mathbb{R}$ satisfying: $\lambda_i \in [0,1]$, $\lambda_i = 1$ in $U_i$ and $\lambda_i = 0$ in $W \setminus V_i$. See Figure 5.2

Fix $\epsilon > 0$ and denote by $d_r$ the $C^r$ topology in $C^r(W, M)$. We define inductively a sequence $g_0, g_1, g_2, \cdots, g_k : W \to M$ as follows. First we define $g_0 = A$. For a suitable Morse function $f_1 \in C^r(U_1, \mathbb{R})$ we define $g_1(x)$ as follows:

$$g_1(x) = \left(\phi_1^1(x), \cdots, \phi_{n-1}^1(x), \lambda_1(x) f_1(x) + (1 + \lambda_1(x))(\pi_1 \circ g_0)(x)\right)$$

if $x \in W_1$ and

$$g_1(x) = g_0(x)$$

if $x \in W \setminus W_1$.

It follows that

$$d_r(g_1, g_0) \leq d_r(g_1/W_1, g_0/W_1)(\Sigma_{j=1}^r K_j d_j(\lambda_1, Id/W_1)) \cdot d_r(f_1, \pi_1 \circ g_0),$$

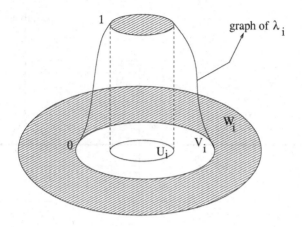

Fig. 5.2

where the constants $K_j$ does not depend on $d_r(f_1, \pi_1 \circ g_0)$. As $\lambda_1$ is fixed we have that $d_r(\lambda_1, Id/W_1)$ does not depend on $d_r(f_1, \pi_1 \circ g_0)$. Hence by Morse Theorem we can choose $f_1$ so that

$$d_r(g_1, g_0) < \epsilon/k.$$

Summarizing $g_1$ satisfies the following properties:

- $d_r(g_1, g_0) < \epsilon/k$;
- $g_1/U_1$ is Morse (because $g_1/U_1 = f_1$).

Replacing $g_0$ by $g_1$ in the above construction we can find $g_2$ such that

- $d_r(g_2, g_1) < \epsilon/k$;
- $g2/(U_1 \cup U_2)$ is Morse.

Repeating the argument we can find the sequence $g_0, g_1, \cdots, g_k$. An element $g_i$ of this satisfies:

- $d_r(g_i, g_{i-1}) < \epsilon/k$;
- $g_i/(U_1 \cup \cdots \cup U_i)$ is Morse.

The last map $g_k$ of the sequence is Morse (because $U_1, \cdots, U_k$ is a cover of $W$). Moreover,

$$d_{C^r}(g_k, A) \leq \Sigma_{i=0}^{k} d_{C^r}(g_i, g_{i+1}) < (\epsilon/k) \cdot k = \epsilon.$$

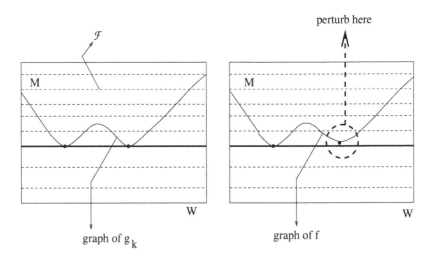

Fig. 5.3

Hence the last map $g_k : W \to M$ is Morse respect to $\mathcal{F}$ and $\epsilon$-close to $A$ in the $C^r$ topology. It remains to choose $f$ close to $g_k$ satisfying the property (2) of the theorem. To modify $g_k$ to obtain $f$ satisfying (1) and (2).

Because the set of Morse function is open we we only have to approximate $g_k$ by a map satisfying the property (2). The last can be attained as in the trivial case described in Figure 5.3 where $M$ is the product foliation $W \times I$ and $\mathcal{F}$ is the trivial foliation $* \times I$. Indeed we only have to perturb around a critical value as indicated in the figure. The theorem is proved.                                                                                                            □

## 5.3    Vector fields on the two-disc

We denote by $D^2$ the 2-dimensional disc in $\mathbb{R}^2$ and by $\mathfrak{X}^1(D^2)$ the set of $C^1$ vector fields in $D^2$ transverse to the boundary $\partial D^2$ of $D^2$. The closure of $B \subset D^2$ is denoted by $\overline{B}$. The orbit of $x \in D^2$ is denoted by $O(x)$. Consider $p \in D^2$, $Y \in \mathfrak{X}^1(D^2)$ and denote by $\omega(p)$ the $\omega$-limit set of $p$. Note that if $\Sigma$ is an interval transverse to $Y$ and $p \in D^2$ is regular, then $\omega(p)$ intersects $\Sigma$ at most once. In particular, a periodic orbit of $Y$ intersects $\Sigma$ once. These facts follow from the trivial topology of the disc $D^2$. A singularity of $Y$ is called *saddle* or *center* according to the two portrait face corresponding to Figure 5.1 (left-hand one for center and right-hand one for saddle). A saddle singularity exhibits two stable separatrices and two unstable separatrices. A *graph* of $Y$ is a connected set $\Gamma$ formed by saddles and saddle's separatrices in a way that if $s \in \Gamma$ is a saddle then $\Gamma$ contains at least one stable separatrix and one unstable separatrix of $s$.

**Theorem 5.4.** *Let* $Y \in \mathfrak{X}^1(D^2)$ *be such that* $Y$ *is transverse to* $\partial D^2$ *and* $\mathrm{sing}(Y)$ *is a finite set formed by centers and saddles. Suppose that* $Y$ *has no saddle-connections. Then, there is* $x \in D^2$ *such that:*

1) $\overline{O(x)}$ *is a closed curve.*
2) *There is an interval* $\delta$ *transverse to* $Y$ *with the following properties:*
   2.1) $\delta \cap \overline{O(x)} = \delta \cap O(x) = \{x\}$.
   2.2) *The first return map* $f \colon \mathrm{Dom}(f) \subseteq \delta \to \delta$ *induced by* $Y$ *in* $\delta$ *satisfies that:* $f = \mathrm{Id}$ *in a connected component of* $\delta \setminus \{x\}$ *and* $f \neq \mathrm{Id}$ *in any neighborhood of* $x$ *in* $\delta$.

**Proof.** Because $Y$ has no saddle-connection the graphs of $Y$ are as in Figure 5.4. Clearly the complement $D^2 \setminus \Gamma$ of a compact invariant set $\Gamma$ equals to either a periodic orbit or a graph contains at least one connected components disjoint from $\partial D^2$. The union of such connected components will be denoted by $R(\Gamma)$. We define an order $<$ on the set formed by periodic orbits and graphs of $Y$ by setting:

$$\Gamma_1 < \Gamma_2 \Leftrightarrow R(\Gamma_1) \subseteq R(\Gamma_2).$$

A *limit cycle* of $Y$ will a compact invariant set $L$ with regular orbits of $Y$ equals to $\omega(p)$ for some $p \notin L$. It is easy to prove that a limit cycle $L$ is either a periodic orbit or a graph. Hence the order $<$ is well defined on the set of limit cycles of $Y$.

**Lemma 5.2.** *If* $\Gamma_1 > \Gamma_2 > \ldots$ *is a decreasing sequence of limit cycles of* $Y$, *then* $\Gamma_\infty = \partial \left( \bigcap_{n=1}^{\infty} R(\Gamma_n) \right)$ *is either a periodic orbit or a graph of* $Y$.

**Proof.** Since $Y$ is transverse to $\partial D^2$ we can assume that $Y$ points inward on the boundary $\partial D^2$ of the disc $D^2$. Clearly, $Y$ has finitely many graphs as it has finitely many singularities. Hence we can assume that $\Gamma_n$ is a periodic orbit, $\forall n$. So, $R(\Gamma_n)$ is a disc and $\partial R(\Gamma_n) = \Gamma_n$, $\forall n$. There is at least one regular point in $\Gamma_\infty$ because if $s \in \Gamma_\infty \cap \mathrm{sing}(Y)$, then $s$ must be saddle and so at least one of the separatrices of $s$ is accumulated by $\Gamma_n$.

First we observe that $\Gamma_\infty$ cannot contain periodic orbits unless it is a periodic orbit. Indeed this follows from Lemma 4.2 but we give a direct

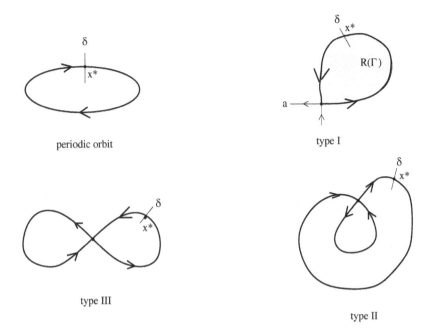

Fig. 5.4

proof here. Suppose that $\Gamma_\infty$ contains a periodic orbit $\alpha$. Pick $x \in \alpha$ and let $\Sigma_x$ be a transverse of $Y$ containing $x$. Clearly for all $n \in \mathbb{N}$ large the set $\Gamma_n \cap \Sigma_x$ consists of a single point $x_n$ such that $x_n \to x$ as $n \to \infty$. Let $A_n$ be the annulus in $D^2$ formed by $\Gamma_n$ and $\alpha$. Since $x_n \to x$ the Tubular Flow-Box theorem implies that $\Gamma_n \to \alpha$ in the Hausdorff topology proving $\Gamma_\infty = \alpha$ as desired.

Second we observe that $\Gamma_\infty$ cannot contain a graph unless it is a graph. The proof of this is similar to the previous proof. Indeed, let $\Gamma \subset \Gamma_\infty$ be a graph. Hence $\Gamma$ is one of the three types of graphs in Figure 5.4. If $\Gamma$ is type II or III then previous argument shows that $\Gamma_\infty = \Gamma$ and we are done. Otherwise $\Gamma$ is type I. In this case $\Gamma$ does not contain an unstable separatrix $a$ (say). By the Poincaré Bendixson Theorem we have that $\omega(a)$ is either a singularity or a graph or a periodic orbit. In the later case we have that $\Gamma_\infty$ is a periodic orbit, a contradiction because it contains the graph $\Gamma$. In the former case we have that $\Gamma$ is one of the graphs in the bottom figures ($Y$ has no saddle connections) again a contradiction. Hence $\omega(p')$ must be a graph which we still denoted by $\Gamma$. $\Gamma$ cannot be of type

I for, otherwise, the limit cycle sequence $\Gamma_n$ must be contained in $R(\Gamma)$ a contradiction since $\Gamma_n$ is decreasing. This proves that $\Gamma$ is type II or III and we are done.

Now, fix $p \in \Gamma_\infty$ regular. Clearly $\Gamma_\infty$ is invariant and so $\omega(p) \subset \Gamma_\infty$. Poincaré-Bendixson implies that $\omega(p)$ is either a singularity or a periodic orbit or a graph. In the last two cases we have that $\Gamma_\infty$ contains either a periodic orbit or a graph. Hence $\Gamma_\infty$ is either a periodic orbit or a graph and we are done. We conclude that $\omega(p)$ is a singularity. Analogously $\alpha(p)$ can be assumed to be a singularity $Y$ because it points inward $D^2$ from $\partial D^2$. Because $Y$ has no saddle connections we have that the closure $\overline{O(p)}$ of the orbit $O(P)$ is a graph of $Y$. This graph is evidently contained in $\Gamma_\infty$. We conclude that $\Gamma_\infty$ contains a graph and we are done. This proves the lemma. □

Let us finish the proof of Theorem 5.4. Consider the set $\mathcal{R}$ of all compact invariant sets $\Gamma_\infty$ of $Y$ of the form

$$\Gamma_\infty = \partial \left( \bigcap_{n=0}^{\infty} R(\Gamma_n) \right),$$

for some decreasing sequence of limit cycles $\Gamma_n$ of $Y$. Lemma 5.2 implies that the order $<$ is defined in $\mathcal{R}$. Lemma 5.2 also implies that any decreasing sequence $\Gamma_\infty^1 > \Gamma_\infty^2 > \cdots$ in $\mathcal{R}$ has an infimum in $\mathcal{R}$. The Zorn Lemma implies that there is a minimal element $\Gamma^* = \Gamma_\infty^*$ in $\mathcal{R}$. By Lemma 5.2 we have that $\Gamma^*$ is either a periodic orbit or a graph. In any case we choose $x^* \in \Gamma^*$ as indicated in Figure 5.4. Observe that the closure $\overline{O(x^*)}$ of the orbit $O(x^*)$ is a closed curve. Choose a transverse $\delta$ containing $x^*$ as indicated in the figure. Let $f : \mathrm{Dom}(f) \subset \delta \to \delta$ be the return map induced by $Y$ in $\delta$. Because $\Gamma^*$ is accumulated by limit cycles of $Y$ one has that $f \neq Id$ in any neighborhood of $x^*$ in $\delta$ (for such limit cycles must intersect $\delta$). On the other hand, consider the connected component $c$ of $\delta - \{x^*\}$ contained in $\Gamma^*$. Because the number of graphs of $Y$ is finite, we can assume by shrinking $\delta$ if necessary that $\delta$ does not intersect any graph of $Y$. In particular $c$ does not intersect any graph of $Y$. Because $c \subset R(\Gamma^*)$ we conclude that the orbit of any point in $c$ is periodic. Hence $f = Id$ in $c$. This proves that $x = x^*$ satisfies the properties (1),(2) of the theorem. □

## 5.4   Proof of Haefliger's theorem

Let $\mathcal{F}$ be a codimension one $C^2$ foliation with a null-homotopic closed transverse $\gamma$. Since $\gamma$ is null-homotopic, there is a $C^\infty$ map $A : D^2 \to M$

such that $A(\partial(D^2)) = \gamma$. By Theorem 5.3 we can assume that $A = f$ is in general position with respect to $\mathcal{F}$. Let $\mathcal{F}^* = f^*(\mathcal{F})$ be the foliation in $D^2$ induced by $f$. By definition a leaf $L$ of $\mathcal{F}^*$ is $f^{-1}$(connected component of $F \cap f(D^2)$) for some leaf $F$ of $\mathcal{F}$. Note that $\mathcal{F}^*$ is a singular foliation of class $C^2$ in $D^2$ and a singularity of $\mathcal{F}^*$ is either a center or a saddle. Clearly $\mathcal{F}^*$ is $C^2$ orientable close to the singularities. Far from the singularities we have that $\mathcal{F}^*$ is $C^2$ locally orientable by the Tubular Flow-Box theorem. We conclude that $\mathcal{F}^*$ is $C^2$ locally orientable. By the results in Section 2.3 we have that $\mathcal{F}^*$ is $C^2$ orientable, *i.e.*, there is a $C^2$ vector field $Y$ in $D^2$ tangent to $\mathcal{F}^*$. Note that $Y$ is transverse to $\partial D^2$ and contains a finite number of singularities all of them being centers or saddles. Moreover, $Y$ has no saddle connections by Theorem 5.3-(2). It follows from Theorem 5.4 that there is $x \in D^2$ and a transverse $\delta$ satisfying the conclusions (1)-(2) in that theorem. In particular, if $x_0 = f(x)$ then $c = f(\overline{O(x)})$ is a closed curve contained in the leaf $F = \mathcal{F}_{x_0}$ of $\mathcal{F}$ and $\Sigma = f(\delta)$ is a transverse segment of $\mathcal{F}$ intersecting $c$. The holonomy of $c$ is conjugated to the return map $f\colon \mathrm{Dom}(f) \subset \delta \to \delta$ induced by $Y$ in $\delta$. One can see that $c$ and $F$ satisfy the properties (1)-(2) of Definition 5.1 by using the property (2.2) in Theorem 5.4. We conclude that $F$ is a leaf with one-sided holonomy of $\mathcal{F}$ proving the theorem.                                                   □

Exercise 5.4.1. Let $\mathcal{F}$ be a codimension one foliation on $S^3$ with a compact leaf $L \in \mathcal{F}$ homologous to zero. Show that $L$ is the torus.

# Chapter 6

# Novikov's compact leaf theorem

## 6.1 Statement

The search for codimension one foliations on the three sphere $S^3$ was ended by the introduction of the Reeb foliation described in Example 1.3. Nevertheless, this example exhibits some particularities, suggesting some of them are part of a global phenomena. One of them is the existence of a torus leaf. The other is the existence of a one-sided holonomy, though this was proved in Haefliger's theorem. Regarding the existence of a compact (torus) leaf, this also part of a global phenomena. Indeed, in this chapter we shall prove the celebrated Novikov's compact leaf theorem.

**Theorem 6.1 (Novikov's compact leaf theorem).** *Codimension one $C^2$ foliations on compact 3-manifolds with finite fundamental group have compact leaves.*

The proof of this theorem given here is the one in [Haefliger (1967)] "Séminaire Bourbaki 20e année, 1967-68, Num. 339, p. 433-444". That proof is based on the following definition.

**Definition 6.1 (vanishing cycle).** Let $\mathcal{F}$ be a $C^1$ codimension one foliation in a manifold $M$. A *vanishing cycle* of $\mathcal{F}$ is a $C^1$ map $f \colon S^1 \times [0, \epsilon] \to M$ (for some $\epsilon > 0$) such that if we denote $f_t(x) = f^x(t) = f(x, t)$, $\forall (x, t) \in S^1 \times [0, \epsilon]$, then the following properties hold:

(1) $f_t(S^1)$ is a closed curve contained in a leaf $A(t)$ of $\mathcal{F}$, $\forall t$;
(2) $f_t(S^1)$ is null homotopic in $A(t)$ if and only if $t > 0$;
(3) $f^x([0, \epsilon])$ is transverse to $\mathcal{F}$, $\forall x$.

**Example 6.1.** Let $\mathcal{F}$ be the Reeb foliation in $S^3$ and let $T$ be the compact leaf of $\mathcal{F}$. Any generator of $\pi(T)$ is represented by a curve contained in (the image of) a vanishing cycle of $\mathcal{F}$.

**Example 6.2.** A torus fibration over $S^1$ gives an example of a foliation with compact leaves having no vanishing cycles.

The proof of Novikov's compact leaf theorem is a direct consequence of the following two preliminary results.

**Theorem 6.2 (Auxiliary theorem I).** *Codimension one $C^2$ foliations on compact 3-manifolds with finite fundamental group have vanishing cycles.*

**Theorem 6.3 (Auxiliary theorem II).** *Codimension one $C^1$ transversely orientable foliations with vanishing cycles on compact 3-manifolds have compact leaves.*

**Proof of Novikov's compact leaf theorem.** Let $\mathcal{F}$ be a codimension one $C^2$ foliation on a compact 3-manifold $M$. Let $P : \hat{M} \to M$ be a finite cover of $M$ such that the lift $\hat{\mathcal{F}}$ of $\mathcal{F}$ is transversely orientable. $\pi_1(\hat{M})$ is finite since $\pi_1(\hat{M}) < \pi_1(M)$ and $\pi_1(M)$ is finite. By Auxiliary theorem I we have that $\hat{\mathcal{F}}$ has a vanishing cycle. Hence $\hat{\mathcal{F}}$ has a compact leaf $\hat{F}$ by Auxiliary theorem II. Then $F = P(\hat{F})$ is a compact leaf of $\mathcal{F}$ proving the result. $\qquad\square$

## 6.2   Proof of Auxiliary theorem I

Let $\mathcal{F}$ be a codimension one $C^2$ foliation on a compact 3-manifold $M$. *It is easy to prove that $\mathcal{F}$ has a closed transverse $\gamma$.* In fact, by passing to a finite cover we can assume that $\mathcal{F}$ is transversely orientable, and so, it has a transverse vector field $X$. Because $M$ is compact we have that $X$ has a non-wandering point $x$. Hence there is a piece of orbit of $X$ which starts and finishes close to $x$. By modifying a bit such a piece of orbit nearby $x$ we can construct a closed transverse of $\mathcal{F}$ containing $x$. This proves the result.

Next we assume that $\pi_1(M)$ is finite and let $\gamma$ be a closed transverse of $\mathcal{F}$. Because $\pi_1(M)$ is finite we have that there is $n \in \mathbb{N}$ such that the curve $\gamma^n$ represent a closed null homotopic transverse of $\mathcal{F}$. Without loss of generality we can assume that $n = 1$. Now we proceed as in the proof of Haefliger's theorem: Because $\gamma$ is null-homotopic one has that there is a $C^\infty$

map $A : D^2 \to M$ such that $A(\partial(D^2)) = \gamma$. By Theorem 5.3 we can assume that $A = f$ is in general position with respect to $\mathcal{F}$. Let $\mathcal{F}^* = f^*(\mathcal{F})$ be the foliation in $D^2$ induced by $f$. Note that $\mathcal{F}^*$ is a singular foliation of class $C^2$ in $D^2$ and a singularity of $\mathcal{F}^*$ is either a center or a saddle. Clearly $\mathcal{F}^*$ is $C^2$ orientable close to the singularities. Far from the singularities we have that $\mathcal{F}^*$ is $C^2$ locally orientable by the Tubular Flow-Box Theorem. Hence $\mathcal{F}^*$ is $C^2$ locally orientable. By the last example of Section 2.2 we have that $\mathcal{F}^*$ is $C^2$ orientable, *i.e.*, there is a $C^2$ vector field $Y$ in $D^2$ tangent to $\mathcal{F}^*$. Note that $Y$ is transverse to $\partial D^2$ and contains a finite number of singularities all of them being centers or saddles. Moreover, $Y$ has no saddle connections by Theorem 5.3-(2). It follows from Theorem 5.4 that there is $x \in D^2$ and a transverse $\delta$ satisfying the conclusions (1)-(2) in that theorem. Let $c_0$ be the closed curve $c_0 = f(\overline{O(x)})$. Then $c_0 \subset \mathcal{F}_{x_0}$ where $x_0 = f(x)$. Note that $c_0$ is *not* null-homotopic in $\mathcal{F}_{x_0}$ because its holonomy map is not the identity in any neighborhood of $x_0$. Moreover, the closed curve $\beta = \overline{O(x)}$ is either a periodic orbit or the closure of a homoclinic loop of $Y$. These properties motivate us to define *cycle* as a closed curve $\beta$ in $D^2$ which is either a periodic orbit or the closure of a homoclinic loop of $Y$ such that $f(\beta)$ is *not* null homotopic in the leaf of $\mathcal{F}$ containing it. As before every cycle $\beta$ of $Y$ bounds a region $R(\beta)$ which does not intersect $\partial D^2$. We define the order $<$ in the set of cycles of $Y$ by setting $\beta_1 < \beta_2$ if and only if $R(\beta_1) \subset R(\beta_2)$.

**Lemma 6.1.** *Let $\beta_1 > \beta_2 > \cdots$ be a decreasing sequence of cycles of $Y$. Then there is a cycle $\beta_\infty$ of $Y$ such that $\beta_n > \beta_\infty$ for all $n \in \mathbb{N}$.*

**Proof.** Because the number of homoclinic loops of $Y$ is finite we can assume that $\beta_n$ is a periodic orbit and that $R(\beta_n)$ is a disc with boundary $\beta_n$ for all $n$. The sequence $R(\beta_n)$ is a decreasing sequence of compact sets in $D^2$. Hence

$$\bigcap_{n=1}^{\infty} R(\beta_n)$$

is a non-empty compact set whose boundary will be denote by $\beta$. It is clear that $\beta \neq \emptyset$. Moreover, there is $p \in \beta$ regular because the singularities of $Y$ are centers or saddles (no periodic orbit close to a center of $Y$ can be a cycle of $Y$). By Poincaré-Bendixson we have that $\omega(p)$ is either a singularity or a periodic orbit or a graph. In the later two cases we have that $\beta$ contains either a periodic orbit or a graph. Hence, as in the proof of Lemma 5.2 in Section 5.3, we have that $\beta$ itself is either a periodic orbit or a graph of type II or III (see Figure 5.4). If $\beta$ were a periodic orbit with $f(\beta)$ null homotopic in its leaf then $f(\beta_n)$ would be null homotopic in its leaf for all $n$ large, a contradiction. Hence if $\beta$ is periodic then $\beta_\infty = \beta$ is the desired cycle. Now suppose that $\beta$ is a graph of type II or III. The fact that $\beta$ is surrounded by cycles of $Y$ implies that $f(\beta)$ is not null homotopic in its leaf. Hence one of the two homoclinic loops forming $\beta$ (say $\beta'$) satisfies that $f(\beta')$ is not null homotopic in its leaf. Then, $\beta_\infty = \beta'$ is the desired cycle. To finish we assume that $\omega(p)$ is a singularity. In a similar way we can assume that $\alpha(p)$ is a singularity. Hence $\overline{O(p)}$ is a graph contained in $\beta$. As previously remarked this implies the existence of the desired cycle $\beta_\infty$ and the proof follows.                                                                                    □

Let us finish the proof of Auxiliary theorem I. By the previous lemma and the Zorn lemma we have that there is a cycle $\beta_\infty$ of $Y$ which is minimal for the order $<$. By the definition of cycle we have that $f(\beta_\infty)$ is not null homotopic in its leaf. Choose a regular point $x \in \beta_\infty$ and let $\delta$ be a transverse of $\mathcal{F}^*$ inside $R(\beta_\infty)$ containing $x$ in its boundary. In order to simplify the notation we shall assume that $\delta = [0,1]$ with $x \approx 0$. By Poincaré-Bendixson, because the number of graphs of $Y$ is finite, we have that all orbit of $Y$ starting at $y \in \delta \setminus \{x\}$ is periodic with period $t_y$. Hence the set

$$\mathcal{A} = \overline{O(x)} \cup \{Y_{[0,t_y]} : y \in \delta \setminus \{x\}\}$$

is diffeomorphic to the annulus $S^1 \times [0,1]$. Put a parametrization $P :$ $S^1 \times [0,1] \to \mathcal{A}$ such that: $P_0(S^1) = \overline{O(x)}$; $P_y(S^1) = Y_{[0,t_y]}$ for all $y \in \delta \setminus \{x\}$; and $P^\theta([0,1])$ is transverse to $\mathcal{F}^*$ for all $\theta \in S^1$. Consider the map $g = f \circ P$. Hence $g : S^1 \times [0,1] \to M$ is a limit cycle of $\mathcal{F}$. In fact, $g$ is $C^1$ since both $P$ and $f$ are. In addition, $g_0(S^1) = f(\overline{O(x)}) \subset \mathcal{F}_{f(x)}$ is not null homotopic in $A(0) = \mathcal{F}_{f(x)}$ because $\beta_\infty$ is a cycle of $Y$. Also for $y \neq 0$ we have that $g_y(S^1) = f(a(y))$ is null homotopic in its leaf since $\beta_\infty$ is minimal with respect to the order $<$. Because $P^\theta([0,1])$ is transverse to $\mathcal{F}^*$ for all $\theta \in S^1$ one has that $g^\theta(S^1) = f(P^\theta([0,1]))$ is transverse to $\mathcal{F}$ for all $\theta \in S^1$. This proves the theorem.                                                                                    □

## 6.3   Proof of Auxiliary theorem II

The proof uses the following holonomy lemma. First, we use introduce short definitions. Given a codimension one foliation $\mathcal{F}$ we say that a vector field $X$ is *normal for $\mathcal{F}$* if the trajectories of $X$ are everywhere transverse to

$\mathcal{F}$. Clearly a codimension one foliation has a normal vector field if and only if it is transversely orientable. If $\mathcal{F}$ is a transversely orientable foliation in a manifold $M$ we say that a curve $c \subset M$ is *normal to* $\mathcal{F}$ if $c$ is transverse to $\mathcal{F}$ and contained in a solution curve of the normal vector field associated to $\mathcal{F}$. Given a leaf $A$ of $\mathcal{F}$, a compact set $K$ and a $C^1$ map $g : K \to A$ we say that $g$ *has a normal extension* if there are $\epsilon > 0$ and a $C^1$ map $G : K \times [0, \epsilon]) \to M$ such that:

(1) $G_0/K = g$;
(2) $G_t(K) \subset A(t)$ for some leaf $A(t)$ of $\mathcal{F}$ with $A(0) = A$;
(3) $\forall x \in K$ the curve $G^x([0, \epsilon])$ is normal to $\mathcal{F}$.

**Lemma 6.2 (Holonomy Lemma).** *Let $\mathcal{F}$ a codimension one transversely orientable foliation, $A$ be a leaf of $\mathcal{F}$ and $K$ be compact set. If $g : K \to A$ is a $C^1$ map homotopic to constant in $A$, then $g$ has a normal extension.*

**Proof.** Let $g : K \to A$ of as in the statement and denote by $X$ the normal vector field of $\mathcal{F}$. For all $x \in K$ define the normal curve

$$\Sigma_x = \{X_s(g(x)) : s \in [0, 1]\},$$

where as usual $X_t$ denotes the flow of $X$ and $X_B(A) = \{X_s(z) : (s, z) \in B \times A\}$. By hypothesis $g(K)$ is a compact null homotopic subset of $A$, and so, it is contained is disc $D \subset A$. Fix $x_0 \in K$ and define

$$G(x, t) = f(X_t(g(x)),$$

where $f = f_x \colon \text{Dom}(f) \subset \Sigma_{x_0} \to \Sigma_x$ is the holonomy induced by a curve $\gamma_x \subset D$ joining $g(x_0)$ to $g(x)$. Since $D$ is contractible we have $G(x, t)$ does not depend on the chosen curve $\gamma_x$. Moreover, since $g(K)$ is compact, we can assume that there is $\epsilon > 0$ such that $G(x, t)$ is defined for every $(x, t) \in K \times [0, \epsilon]$. Let us prove that the map $G : K \times [0, \epsilon] \to M$ so obtained is a normal extension of $g$. It is clear that $G$ is $C^1$ since $g$ is. First we prove that $G_0/K = g$. In fact, if $x \in K$ then $G_0(x) = G(x, 0) = f(X_0(g(x))) = f(g(x)) = g(x)$ by the definition of holonomy and $X_0 = \text{Id}$. Second we prove that $G_t(K) \subset A(t)$ for some leaf $A(t)$ of $\mathcal{F}$ with $A(0) = A$. In fact, it is clear that $A(0) = A$. Next we observe that $G_t(x) = G(x, t) = f(X_t(g(x))) \in \mathcal{F}_{X_t(g(x))}$. Since $\mathcal{F}_{X_t(g(x))} = \mathcal{F}_{X_t(g(x_0))}$ by definition of holonomy we have that $A(t) := \mathcal{F}_{X_t(g(x_0))}$ works. Third we prove that $G^x([0, \epsilon])$ is normal to $\mathcal{F}$. In fact, $G^x(t) = G(x, t) = f(X_t(g(x))) \in \Sigma_x$ which is a solution curve of $X$. The lemma follows. $\qquad \square$

Hereafter we let $\mathcal{F}$ be a codimension one $C^1$ transversely orientable foliation with a vanishing cycle $f\colon S^1 \times [0, \epsilon] \to M$ on a compact 3-manifold $M$. We denote by $X$ the vector field in $M$ transverse to $\mathcal{F}$ and by $X_t$ the flow generated by $X$. This vector field exists because $\mathcal{F}$ is transversely orientable. For simplicity we shall assume that $\epsilon = 1$. A $C^1$ curve $\alpha$ in a leaf of $\mathcal{F}$ is in *general position* whenever $\#\alpha^{-1}(p) \leq 2$ for all $p \in M$ and if $x, y \in \mathrm{Dom}(\alpha)$ are different points with $\alpha(x) = \alpha(y)$, then $\alpha'(x)$ and $\alpha'(y)$ are not parallel (see Figure 6.1).

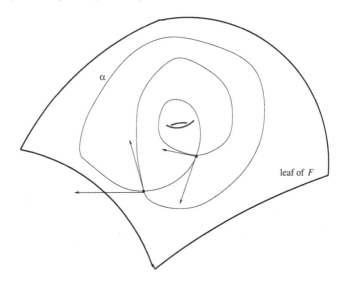

Fig. 6.1    General position curve for $\mathcal{F}$.

**Lemma 6.3.** *We can suppose that the vanishing cycle $f\colon S^1 \times [0, 1] \to M$ of $\mathcal{F}$ satisfies the following additional properties:*

(4) $f_0(S^1)$ *is in general position in $A(0)$.*
(5) $f^x([0, 1])$ *is normal to $\mathcal{F}$.*
(6) *If $x, y \in S^1$ and $f^x(0) \neq f^y(0)$, then $f^x([0, 1]) \cap f^y([0, 1]) = \emptyset$.*

**Proof.** Clearly (5) implies (6) by standard arguments from Ordinary differential equations. By moving a bit $f_x(S^1)$ we can assume (4). To assume (5) it suffices to project $f^x([0, 1])$ to the solution curve of $X$ passing through $f^x(0)$ via holonomy. This is done as follows (see Figure 6.2): For $x \in S^1$ we define $\Sigma = \{X_s(f^x(0)); s \in [0, 1]\}$ which is the solution curve of $X$ passing through $f^x(0)$. We define $\Sigma' = f^x([0,1])$. By the definition of vanishing cycle the curve one has $\Sigma' \pitchfork \mathcal{F}$. As $X \pitchfork \mathcal{F}$, we have $\Sigma \pitchfork \mathcal{F}$. Note that $x \in \Sigma \cap \Sigma'$. Hence there is a holonomy map $g : \text{Dom}(g) \subset \Sigma' \to \Sigma$. Define $f^*(x,t) = g(f^x(t))$. By compactness we can assume that $f^*(x,t)$ is defined in $S^1 \times [0,\epsilon]$ for some $\epsilon > 0$. We shall assume that $\epsilon = 1$ for simplicity. Let us prove that $f^*$ is a vanishing cycle of $\mathcal{F}$. First note that $(f^*)_0(x) = g(f^x(0)) = g(x) = x$ (by the definition of $g$) and so $(f^*)_0(S^1) = f_0(S^1)$. The last implies that $f_0(S^1) \subset A(0)$ is not null-homotopic in $A(0)$. Moreover, $(f^*)_t(x) = f^*(x,t) = g(f^x(t)) \in \mathcal{F}_{f^x(t)} = A(t)$ by the definition of holonomy. Hence $f_t^*(S^1) \subseteq A(t)$, $\forall t$. Now we prove that $F_t^*(S^1)$ is null homotopic in $A(t)$ for all $t > 0$ small. In fact note that $f_t^*(x) = g(f^x(t)) \underset{t \to 0^+}{\longrightarrow} g(f^x(0)) = f_0(x)$ and $f_t(x) \underset{t \to 0^+}{\longrightarrow} f_0(x)$. Hence $d(f_t^*(x), f_t(x) \le d(f_t^*(x), f_0(x)) + d(f_0(x), f_t(x)) \underset{t \to 0^+}{\longrightarrow} 0$. Thus, $f_t^*(S^1)$ is $C^0$-close to $f_t(S^1)$ in $A(t)$. Then, $f_t^*(S^1) \simeq 0$ em $A(t)$ as desired. The proof follows. $\qquad\square$

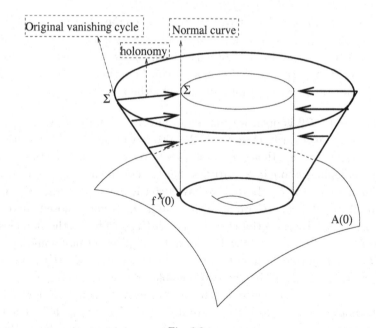

Fig. 6.2

**Lemma 6.4.** *We can assume that the vanishing cycle* $f: S^1 \times [0,1] \to M$ *of* $\mathcal{F}$ *satisfies the following additional property:*

(7) *The lift* $\hat{f}_t : S^1 \to \hat{A}(t)$ *of* $f_t : S^1 \to A(t)$ *to the universal cover* $\pi(t): \hat{A}(t) \to A(t)$ *of* $A(t)$ *is a* simple closed curve, $\forall t > 0$.

**Proof.** Define $R = \{$couples $(x,y)$ of different points of $S^1$ such that $\hat{f}_t(x) = \hat{f}_t(y)$ for some $t > 0\}$, and $r_t = \#R$. We have that $r < \infty$. In fact, for all $t$ we define $B_t = \{p \in f_t(S^1) : \#f^{-1}(p) = 2\}$, $b_t = \#B_t$. $b_0 < \infty$ because $f_0$ is in general position. Note that $f_t \to f_0$ in the $C^1$ topology as $t \to 0^+$ and, because $f_0$ is in general position we have $b_t = b_0$ for all $t \approx 0$. Hence the map $t \to b_t$ is the constant map $t \to b_0$. As $r \leq b_0$ the result follows. On the other hand, $R \neq \emptyset$ for otherwise we are done. Pick $(x,y) \in R$ and define $U = \{t > 0 : \hat{f}_t(x) = \hat{f}_t(y)\}$, $K = \{t \leq 0 : f_t(x) = f_t(y)\}$. The following properties hold:

- $K$ is compact (because $f^x$ and $f^y$ are continuous).
- $U \neq \emptyset$ (because $(x,y) \in R$).
- $U \subset K$ (because $\hat{f}_t(x) = \hat{f}_t(y) \Rightarrow f(t(x) = f_t(y))$).
- $U$ is open: Fix $t$ such that $\hat{f}_t(x) = \hat{f}_t(y)$. Then the curve $(\pi(t) \circ \hat{f}_t)/[x,y]$ is closed, where $[x,y]$ is a suitable arc in $S^1$ joining $x,y$. This curve is null homotopic in $A(t)$ because its lift $\hat{f}_t/[x,y]$ in $\hat{A}(t)$ is a closed curve. By the Holonomy Lemma we conclude that $(\pi(s) \circ \hat{f}_s)/[x,y]$ is null homotopic in $A(s)$ for $s \approx t$ proving the result.

Because $U \neq \emptyset$ is open we can fix an interval $(t', t'']$ in $U$ with $t'' \notin U$ and $t''$ being arbitrarily close to 0. We also have that $t' \in K$, because $K$ is closed and $U \subset K$. Hence $f_{t'}(x) = f_{t'}(y)$. We claim that one of the two arcs $[x,y]$ joining $x$ to $y$ in $S^1$ satisfies that the closed curve $f_{t'}([x,y])$ is *not* null homotopic in $A(t')$. Indeed we have two cases, namely either $t' > 0$ or $t' = 0$. If $t' > 0$ as $t' \notin U$ we have that $\hat{f}_{t'}([x,y])$ is not a closed curve for each arc $[x,y]$. Because the closed curve $f_{t'}([x,y])$ lifts to the non-closed curve $\hat{f}_{t'}([x,y])$ in $\hat{A}(t')$ we conclude that $f_{t'}([x,y])$ is not null homotopic in $A(t)$ for all $[x,y]$ (recall that $\hat{A}(t')$ is the universal cover of $A(t')$). If $t' = 0$ then we let $[x,y], [x,y]'$ be the two possible arcs in $S^1$ joining $x$ to $y$. If both $f_0([x,y]), f_0([x,y]')$ were null homotopic in $A(0)$ then $f_0(S^1)$ would be null homotopic in $A(0)$ as it is the product of $f_0([x,y]), f_0([x,y]')$. The last contradicts the definition of vanishing cycle. This proves the claim. Hence we can assume that $f_{t'}([x,y])$ is not null-homotopic in $A(t')$. We note that the closed curve $\hat{f}_{t'}([x,y])$ has less than $r$ multiple points. Moreover the restriction $f/(S^1 \times [t', t''])$ : $S^1 \times [t', t''] \to M$ is a vanishing cycle of $\mathcal{F}$ with $A(t')$ *close to* $A(0)$. Replacing $f$ by $f/(S^1 \times [t', t''])$ we have less than $r$ multiple points for the new vanishing cycle. Repeating the process we obtain the result. $\qquad \square$

**Lemma 6.5.** *Let* $f \colon S^1 \times [0, 1] \to M$ *be a vanishing cycle of* $\mathcal{F}$ *satisfying the properties* (4)-(6) *of* Lemma 6.3 *and* (7) *of* Lemma 6.4. *Then, there is an immersion* $F \colon D^2 \times (0, 1] \to M$ *satisfying the following properties:*

(i) $F_t|_{\partial D^2} = f_t$, $\forall t$.
(ii) $F(D^2 \times t) \subset A(t)$, $\forall t$.
(iii) $F^x((0, 1])$ *is normal to* $\mathcal{F}$, $\forall x$.
(iv) *If* $U = \{x \in D^2 : \lim_{t \to 0^+} F^x(t) \text{ exists}\}$, *then* $\partial D^2 \subset U$; $U$ *is open; and* $D^2 \setminus U \neq \emptyset$.

**Proof.** Let $\pi(t) : \hat{A}(t) \to A(t)$ be the universal cover of $A(t)$. We have that $\hat{A}(t) = \mathbb{R}^2$ or $S^2$. The last cannot happen for otherwise the Reeb global stability theorem would imply $M = S^2 \times S^1$ and $\mathcal{F}$ is the trivial foliation $S^2 \times *$, a contradiction since $f_0(S^1)$ is not null homotopic in $A(0)$ (see the last exercise in Section 4.3). On the other hand, (7) of Lemma 6.4 says that $\hat{f}_1(S^1)$ is a simple closed curve in $\hat{A}(1)$. By the classical Jordan Theorem we have that there is an embedding $\hat{F} : D^2 \times 1 \to \hat{A}(1)$ with $\hat{F}_1/\partial D^2 = \hat{f}_1$. We define $F : D^2 \times 1 \to A(1)$ by $F = \pi(1) \circ \hat{F}$. Clearly $F$ is an immersion as $\pi(1)$ is a cover and $\hat{F}$ is an embedding. Applying the Holonomy Lemma to $F$ we can extend $F$ to $D^2 \times (t_0, 1]$ for some $t_0 > 0$ satisfying (1)-(3).

We claim that $F$ can be extended to $D^2 \times [t_0, 1]$ still satisfying (1)-(3). In fact, first we show that $\lim_{t \to t_0^+} F(x, t)$ exists for all $x \in D^2$. Because $t_0 > 0$ we have that $\hat{f}_{t_0} : S^1 \to \hat{A}_{t_0}$ is null homotopic. As before there is an embedding $\hat{H}_{t_0} : D^2 \to \hat{A}(t_0)$ with $\hat{H}_{t_0}/\partial D^2 = \hat{f}_{t_0}$. Define $H_{t_0} = \pi(t_0) \circ \hat{H}_{t_0}$. Again by the Holonomy Lemma there are $\delta > 0$ and an immersion $G \colon D^2 \times (t_0 - \delta, t_0 + \delta) \to M$ such that:

   a) $G_t(D^2) \subseteq A(t)$;
   b) $G_t/\partial D^2 = f_t$;
   c) $G^x((t_0 - \delta, t_0 + \delta))$ is normal to $\mathcal{F}$;
   d) $G_{t_0} = H_{t_0}$.

Now we fix $t \in (t_0, t_0 - \delta)$ and consider $D := F_t(D^2)$ and $D^1 := G_t(D^2)$. Both $D$ and $D^1$ are discs contained in $A(t)$ with $\partial D = \partial D^1 = f_t(S^1)$. If $D \neq D^1$ then $A(t)$ would be $S^2$ a contradiction as before by Reed Stability. Hence $D = D^1$ and so $F_t(D^2) = G_t(D^2)$ for all $t \in (t_0, t_0 - \delta)$. If $x \in D^2$ and

$t \in (t_0, t_0 - \delta)$, then $F_t(x) \in F_t(D^2) = G_t(D^2) \Rightarrow F_t(x) = G_t(y(x,t))$ for some $y(x,t) \in D^2$. But $G^{y(x,t)}(t_0 - \delta, t_0 + \delta)$ and $F^x((t_0, 1])$ are both normal to $\mathcal{F}$. Hence $y(x,t) = y(x)$ does not depend on $t$. Thus, $\lim_{t \to t_0^+} F(x,t) = \lim_{t \to t_0^+} G(y(x),t) = G(y(x),t_0) \subseteq A(t_0)$ proving that $\lim_{t \to t_0^+} F(x,t)$ exists $\forall x \in D^2$. To finish the proof of the claim we simply define $H : D^2 \times [t_0, 1] \to M$ by

$$H(x,t) = \begin{cases} \lim_{t \to t_0^+} F(x,t), & \text{if } t = 0, \\ F(x,t), & \text{if } t \neq t_0. \end{cases}$$

Thus $H/D^2 \times (t_0, 1] = F$, $H_t|\partial D^2 = f_t$, $H_t(D^2) \subset A(t)$, $\forall t$ and $H^x([t_0, 1])$ is normal to $\mathcal{F}$. In other words $H$ is an extension of $F$ to $D^2 \times [t_0, 1]$ satisfying (1)-(3). This proves the claim. If $t_0 > 0$ the Holonomy Lemma allow us to extend $F$ to $D^2 \times (t_0 - \delta, 1]$ satisfying (1)-(3). Hence we can assume that there is $F : D^2 \times (0, 1] \to M$ satisfying (1)-(3).

Let us prove that $F$ satisfies the property (4) of the lemma. If $x_0 \in U \Rightarrow \exists y_0 = \lim_{t \to 0^+} F(x_0, t)$. Let $V$ be a tubular flow-box for $X$ around $y_0 \subseteq$ solution curve of $X$. Note that $y_0 \in O_X(x_0)$, the orbit of $x_0$, as $X$ is non-singular ($X \pitchfork \mathcal{F}$). Hence, $X_{t_0}(x_0) = y_0$, for some $t_0 > 0$. By the Tubular Flow-Box Theorem there is a neighborhood $B$ of $x_0$, such that $X_{t_0}(B) \subseteq V$. As $F_1$ is continuous there is a neighborhood $W$ of $x_1$ in $D^2$ with $x_0 = F_1(x_1)$ such that $F_1(W) \subseteq B$. See Figure 6.3.

Let us prove that $\lim_{t \to 0^+} F(x,t)$ exists $\forall x \in W$. In fact, Consider $x \in W$ and $x' = F_1(x)$. Note that the curve $F^{x'}((0,1])$ has finite length for, otherwise, it would exist a first exit point $z$ of $F^{x'}((0,1])$ from $V$. Clearly $z = F^{x'}(t_z)$ for some $t_z \in [0,1]$. But $F^x(t_z) = F_{t_z}(x) \in A(t_z)$, $F^{x'}(t_z) = F_{t_z}(x') \in A(t_z)$. As $z \in \partial V$ and $F^x(t_z) \notin \partial V$ (such a point is close to $y \in \text{Int}(V)$) we conclude by the Mean Value Theorem that $A(t_z)$ and $\mathcal{F}_y$ have an intersection point. This intersection point implies $A(t_z) = \mathcal{F}_y$. Because $\dim \mathcal{F} = 2$ we can assume from the beginning that $\mathcal{F}_y \neq A(t_z)$ a contradiction. Hence $F^{x'}((0,1])$ has finite length $\Rightarrow \exists \lim_{t \to 0^+} F(x',t)$, $\forall x' \in W$. This proves that $U$ is open.

**Warning:** The last argument proves that if $x \in U$ and $y = \lim_{t \to 0^+} F^x(t)$, then there is a neighborhood $W$ of $x$ in $D^2$ such that $\lim_{t \to 0^+} F^{x'}(t) = y'$ exists for all $x' \in W$ and $\mathcal{F}_y = \mathcal{F}_{y'}$.

To see $\partial D^2 \subset U$ it suffices to observe that $F^x(t) = f^x(t)$ for all $x \in$

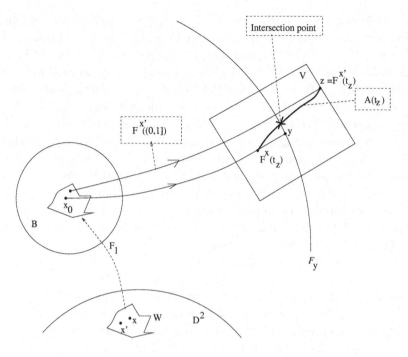

Fig. 6.3

$\partial D^2 = S^1 \Rightarrow \lim_{t\to 0^+} F^x(t) = \lim_{t\to 0^+} f^x(t)$ exists and belongs to $A(0)$ for all $x \in S^1$.

Finally we prove that $D^2 \setminus U \neq \emptyset$. If $D^2 = U$ we would have that $F_0 := \lim_{t\to 0^+} F^x(t)$ would exist for all $x \in D^2$. By the Warning above we would have $F_0(x) \in A(0)$ for all $x \in D^2$. The resulting map $F_0 : D^2 \to A(0)$ yields a continuous extension of $f_0$ to $D^2$, a contradiction since $f_0(S^1)$ is not null homotopic in $A(0)$. This contradiction shows $D^2 \setminus U \neq \emptyset$ and the lemma follows. $\qquad\square$

**Lemma 6.6.** *Let $F$ be the immersion in Lemma 6.5. Then, $\forall \alpha > 0$ there are $0 < t' < t'' < \alpha$ and an embedding $h : D^2 \to int(D^2)$ such that*

$$F(t'', x) = F(t', x), \quad \forall x \in D^2.$$

**Proof.** By Lemma 6.5 there is $y_0 \in D^2 \setminus U$. Hence, the limit $\lim_{t\to 0^+} F^{y_0}(t)$ does not exist. Nevertheless the compactness of $M$ implies that there is a sequence $t_n \to \infty$ such that $F^{y_0}(t_n) \to z$ for some $z \in M$. By using the Tubular Flow-Box Theorem we can assume that $F^{y_0} \in \mathcal{F}_z$ for all

$n$. In addition we can further assume $\mathcal{F}_z \neq A(0)$ because the leaves of $\mathcal{F}$ are two-dimensional. As $F^{y_0}(t_n) \in A(t_n)$ we have $A(t_n) = A(t_m)$ for all $n, m$. Defining $D(t) = F_t(D^2)$ we have $D(t_n) \subset A(t_n) \subset \mathcal{F}_z$, i.e., $D(t_n) \subset \mathcal{F}_z$ for all $n$. Note that $z \in A(t_n)$ $\forall n$ large for, otherwise, it would exist $n_k \to \infty$ with $z \notin D(t_{n_k})$. By hypothesis $F^{y_0}(t_{n_k}) = F_{t_{n_k}}(y_0) \in D(t_{n_k})$ converges to $z$. Because $\partial D(t) = f_t(S^1)$ for all $t$ we conclude that $\exists b_{n_k} \in f_{t_{n_k}}$ sequence converging to $z$. But the distance $dist(b_{n_k}, f_0(S^1))$ goes to 0 as $k \to \infty$. As $f_0(S^1)$ is compact and $b_{n_k} \to z$ we would have $z \in f_0(S^1) \subset A(0)$ yielding $\mathcal{F}_z = A(0)$ a contradiction. This proves that $z \in A(t_n)$ $\forall n$ large. Next we claim that for all $m \in \mathbb{N}$ one has $D(t_m) \subset \mathrm{Int}(D(t_n))$ $\forall n$ large. In fact, note that $\partial D(t_n) = f_{t_n}(S^1)$ hence $\partial D(t_n) \to f_0(S^1)$ uniformly as $n \to \infty$. Clearly we can assume from the beginning that $A(t_n) \neq A(0)$ for all $n$. Hence $f_{t_n}(S^1) \cap A(0) = \emptyset$ for all $n$. It follows that for $m \in \mathbb{N}$ fixed one has $\partial D(t_n) \cap \partial D(t_m) = \emptyset$ for all $n$ large. On the other hand, we can assume $z \in D(t_n)$ for all $n$. From this and $\partial D(t_n) \cap \partial D(t_m) = \emptyset$ one has either $D(t_m) \subset \mathrm{Int}(D(t_n))$ or $D(t_n) \subset \mathrm{Int}(D(t_m))$ for all $n$ large. In the second case we would have $f_0(S^1) \subset D(t_m)$ by taking the limit of the sequence $\partial D(t_n) = f_{t_n}(S^1)$. This would imply $A(t_n) = A(0)$ a contradiction. This contradiction proves $D(t_m) \subset \mathrm{Int}(D(t_n))$ for all $n$ large.

The last claim implies that for $\alpha > 0$ fixed there are $0 < t_n < t_m < \alpha$ such that $D(t_m) \subset D(t_n)$. Choose $t'' = t_m$ and $t' = t_n$. Clearly $A(t') = A(t'')$. To find the embedding $h$ we let $\hat{F}_{t'} : D^2 \to \hat{A}(t')$ and $\hat{F}_{t''} : D^2 \to \hat{A}(t'')$ be the lift to the universal cover. They exist because $D^2$ is contractible. Note that $F_{t''}(D^2) \subset \mathrm{Int}(F_{t'}(D^2))$. Hence for a suitable base point one has $\hat{F}_{t''}(D^2) \subset \mathrm{Int}(\hat{F}_{t'}(D^2))$. As both $\hat{F}_{t'}, \hat{F}_{t''}$ are diffeomorphisms we can define

$$h = (\hat{F}_{t'})^{-1} \circ \hat{F}_{t''} : D^2 \to \mathrm{Int}(D^2).$$

Hence $h$ is an embedding satisfying

$$\hat{F}_{t'}(h(x)) = \hat{F}_{t''}(x), \quad \forall x \in D^2.$$

By composition with the projection $\hat{A}(s) \to A(s)$ for $s = t', t''$ one has the desired property. The lemma follows.    □

**Lemma 6.7.** *Let $f : S^1 \times [0,1] \to M$ be a vanishing cycle of $\mathcal{F}$, for which there is an embedding $F : D^2 \times [0,1] \to M$ satisfying the conclusion of Lemma 6.6. Then, there is in the closed transverse of $\mathcal{F}$ intersecting $A(0)$.*

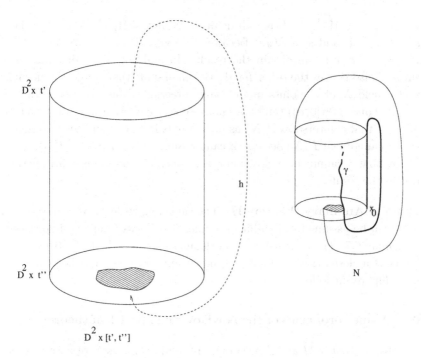

Fig. 6.4

**Proof.** Suppose that there is a closed transverse $\gamma$ of $\mathcal{F}$ intersecting $A(0)$. Modifying a bit $\gamma$ we can assume that there are $x_0 \in S^1$ and $\alpha > 0$ such that

$$f^{x_0}([0, \alpha]) \subset \gamma,$$

and

$$f^x([0, \alpha]) \cap \gamma = \emptyset, \quad \forall x \in S^1 \setminus \{x_0\}.$$

By hypothesis there are $0 < t' < t'' < \alpha$ satisfying the conclusion of Lemma 6.6. Let $h : D^2 \to \text{Int}(D^2)$ be the corresponding embedding. In the cylinder $D^2 \times [t', t'']$ we consider the identification $(x, t'') \approx (h(x), t')$. The manifold $N$ obtained from this identification is depicted in Figure 6.4. Note that $N$ is either a solid torus or a solid Klein bottle depending on whether $h$ preserves or reverses the orientation in $D^2$. In any case we let $\Pi : D^2 \times [t', t''] \to N$ be the quotient map. Denote by $P : N \to M$ the map defined by

$$P(z) = F(x, t),$$

where $(x,t) \in \Pi^{-1}(z)$. $P$ is well defined. In fact, if $\Pi(x,t) = \Pi(y,s)$ then $s = t'', t = t'$ and $x = h(y)$. Because $F(y,s) = F(y,t'') = F(h(y),t') = F(x,t') = F(x,t)$ we obtain the result. In addition $P$ is an immersion since $F$ also is. On the other hand, we can assume that $\gamma$ and the normal vector field $X$ of $\mathcal{F}$ points in the same direction. Now, as $\gamma$ contain the normal curve $f^{x_0}([0, \alpha])$ (and does not intersect any other normal segment) we have that $\gamma$ intersects $P(N)$ as in Figure 6.4. Now it suffices to observe that $\gamma$ cannot exit $P(N)$ because it cannot intersect $P(\Pi(S^1 \times [t', t'']))$. This proves that $\gamma$ cannot be a closed curve, a contradiction. This contradiction proves the result.                                                                        □

**Proof of Auxiliary theorem II:** The vanishing cycle $f \colon S^1 \times [0,1] \to M$ of $\mathcal{F}$ can be assumed to exhibit an immersion $F$ satisfying the hypothesis of Lemma 6.7. The conclusion of this lemma says that the leaf $A(0)$ cannot intersect a closed transverse of $\mathcal{F}$. And this implies that $A(0)$ is a compact leaf. The result follows.                                                                        □

## 6.4    Some corollaries of the Novikov's compact leaf theorem

We observe that if $M$ and $\mathcal{F}$ are orientable and transversely orientable then the quotient manifold $N$ in the proof above is a solid torus. In such a case it can be proved without difficulty that $N$ is a Reeb component of $\mathcal{F}$. This remark is summarized in the following result.

**Theorem 6.4.** *A codimension one transversely orientable $C^2$ foliation on a compact orientable 3-manifold with finite fundamental group has a Reeb component.*

**Corollary 6.1.** *If $\mathcal{F}$ is a codimension one $C^2$ foliation without compact leaves of a compact 3-manifold $M$, then the leaves of $\mathcal{F}$ are $\pi_1$-injectively immersed in $M$.*

**Proof.** Suppose by contradiction that there is a leaf $F$ such that $Ker(\pi_1(F) \to \pi_1(M)) \neq 0$, where $\pi_1(F) \to \pi_1(M)$ is the homomorphism induced by the inclusion $F \to M$. To get the contradiction it suffices by Auxiliary theorem II to prove that $\mathcal{F}$ has a vanishing cycle. For this we proceed as follows. As $Ker(\pi_1(F) \to \pi_1(M)) \neq 0$ there is a curve $\alpha \subset F$ which is null homotopic in $M$ but not in $F$. Because $\alpha$ is null homotopic in $M$ we have that there is a map $f \colon D^2 \to M$ with $\alpha = f(\partial D^2)$. We can assume that $f$ is in general position with respect to $\mathcal{F}$. Hence the induced foliation

$\mathcal{F}^*$ in $D^2$ is a singular foliation tangent to a vector field $Y$. Note that the singularities of $Y$ are either saddles or centers and there is in the saddle connection for $Y$. Clearly the closed curve $\partial D^2$ is a cycle of $Y$. Hence the set of cycles of $Y$ is not empty. By Lemma 6.1 such a set is inductive with respect to the inclusion order. A minimal element in this set produces a vanishing cycle for $\mathcal{F}$ (see the proof of Auxiliary theorem I in Section 6.2). This yields the desired contradiction and the proof follows. $\square$

**Corollary 6.2.** *Let $\mathcal{F}$ be a codimension one $C^2$ foliation without compact leaves of a compact 3-manifold $M$. Then the lift of $\mathcal{F}$ to the universal cover of $M$ is a foliation by planes.*

**Proof.** Let $\hat{\mathcal{F}}$ be the lift of $\mathcal{F}$ to the universal cover $\pi : \hat{M} \to M$ of $M$. Suppose by contradiction that there is a non-simply connected leaf $\hat{F}$ of $\hat{\mathcal{F}}$. Hence there is a closed curve $\hat{c} \subset \hat{F}$ which is not null homotopic in $\hat{F}$. Obviously $\hat{c}$ is null homotopic in $\hat{M}$. Hence the closed curve $c = \pi(\hat{c})$ is null homotopic in $M$. Because $\hat{c}$ is not null homotopic in $\hat{F}$ we have that $c$ is not null homotopic in the leaf $F = \pi(\hat{F})$ of $\mathcal{F}$. This proves that $F$ is *not* $\pi_1$-injectively immersed in $M$. Then $\mathcal{F}$ has a compact leaf by Corollary 6.1, a contradiction. This contradiction proves that all the leaves $\hat{F}$ of $\hat{\mathcal{F}}$ are simply connected. Thus $\hat{F} = \mathbb{R}^2$ or $S^2$. If some $\hat{F}$ is $S^2$ then $\mathcal{F}$ has a compact leaf with finite fundamental group. By Reeb global stability theorem it would follow that all the leaves of $\mathcal{F}$ are compact, a contradiction. This contradiction proves that all the leaves of $\hat{\mathcal{F}}$ are planes as desired. $\square$

**Remark 6.1.** Corollary 6.2 shows that closed 3-manifolds supporting codimension one $C^2$ foliations without compact leaves are *irreducible*, namely every tamely embedded 2-sphere in the manifold bounds a 3-ball. In particular such manifolds are prime, *i.e.*, they are not non-trivial connected sum. We observe that compact 3-manifolds supporting Reebless foliations may be non-irreducible as shown the trivial foliation $\{S^2 \times *\}$ of $S^2 \times S^1$. Nevertheless the 2-sphere bundles over $S^1$ are the solely closed 3-manifolds which are not irreducible and supports Reebless foliations.

**Remark 6.2.** The results in this section hold true for $C^1$ foliations.

Exercise 6.4.1. Let $\mathcal{F}$ be a codimension one foliation on $S^3$ with a compact leaf $L \in \mathcal{F}$ homologous to zero. Show that $L$ is the torus.

# Chapter 7

# Rank of 3-manifolds

The notion of *rank* of a manifold was introduced by J. Milnor, improving original ideas of H. Hopf, in the search of non-homotopic invariants for manifolds.

**Definition 7.1.** Let $M$ be a differentiable manifold. The *rank* of $M$ is the maximum number $k \in \mathbb{N}$ such that there exist continuous vector fields $X_1, \ldots, X_k$ on $M$ with the property that $[X_i, X_j] = 0$, $\forall i, j$ (*i.e.*, the vector fields commute) and $X_1, \ldots, X_k$ being linearly independent at each point of $M$.

The Poincaré-Hopf-Euler Theorem states that any continuous tangent vector field on $S^2$ must have some singularity so that rank $(S^2) = 0$. the following remarkable result is due to E. Lima:

**Theorem 7.1 (Lima's theorem, [Lima (1965)]).** *The rank of the 3-sphere $S^3$ is one.*

Notice that, since a $C^1$ vector field on a *compact* manifold is always complete we may state:

{A compact manifold $M$ has rank $\geq k$} $\Leftrightarrow$ {$M$ admits a locally free action $\varphi \colon \mathbb{R}^k \times M \to M$ of the additive group $(\mathbb{R}^k, +)$}

**Sketch of the proof of Lima's theorem:**

First we observe that rank $(S^3) \geq 1$ as it is easily proved by observing that $X(1, x_2, x_3, x_4) = (-x_2, x_1, -x_4, x_3)$ is tangent to $S^3$ and non-singular (outside of the origin $0 \notin S^3$).

Assume by contradiction that rank $(S^3) \geq 2$. By the above remark there exists a locally free action $\varphi \colon \mathbb{R}^2 \times S^3 \to S^3$. The action generates a codimension one foliation $\mathcal{F}$ (assumed to be $C^2$) on $S^3$. By Novikov's compact leaf theorem $\mathcal{F}$ exhibits some Reeb-component. Thus we are finished once we prove the following:

**Lemma 7.1.** *Given any pair of commuting continuous vector fields $X, Y$ in the solid torus $\overline{D}^2 \times S^1$ such that $X$ and $Y$ are tangent to and linearly independent along $S^1 \times S^1 = \partial(\overline{D}^2 \times S^1)$, then there exists some point $p \in D^2 \times S^1$ where $X$ and $Y$ are linearly dependent.*

**Proof.** The boundary torus $\partial(\overline{D}^2 \times S^1)$ has isotropy group of the form $r \cdot \mathbb{Z} + s \cdot \mathbb{Z}$ for some $r, s \in \mathbb{R}^2 \simeq \mathbb{C}$ with $r/s \notin \mathbb{R}$ so that we may reparameterize $\varphi$ as $\varphi((rt_1, st_2), \cdot)$ $(t_1, t_2) \in \mathbb{R}^2$, in such a way that we may assume

$$X|_{S^1 \times S^1} = -y\frac{\partial}{\partial x} + x\frac{\partial}{\partial y} \quad \text{and} \quad Y|_{S^1 \times S^1} = \frac{\partial}{\partial z}$$

for coordinates $(x, y, z) \in \mathbb{R}^3$ with

$$\overline{D}^2 = \{(x, y, z) \in \mathbb{R}^3; \quad z = 0, \quad x^2 + y^2 \le 1\}.$$

We shall therefore prove that any continuous extension $\tilde{\vec{\varepsilon}}_1, \tilde{\vec{\varepsilon}}_2$ of the vector fields $\vec{\varepsilon}_1 = -y\frac{\partial}{\partial x} + x\frac{\partial}{\partial y}$, $\vec{\varepsilon}_2 = \frac{\partial}{\partial z}$ on $S^1 \times S^1$ to $\overline{D}^2 \times S^1$ must exhibit some point where $\tilde{\vec{\varepsilon}}_1$ and $\tilde{\vec{\varepsilon}}_2$ are linearly dependent. This is done as follows: we may assume that $\tilde{\vec{\varepsilon}}_1$ and $\tilde{\vec{\varepsilon}}_2$ are orthonormal extensions as it is easy to see.

Such an extension may be regarded as a path homotopy $\tilde{a}\colon \partial\overline{D}^2 \times [0, 1] \to G^o_{2,3}$ of the path $a = \tilde{a}(\,\cdot\,, 0)\colon \partial\overline{D}^2 \to G^o_{2,3}$ with a constant; where $G_{2,3}$ is the space of orthonormal oriented pairs of vector on $\mathbb{R}^3$.

By its turn $G^o_{2,3}$ may be identified with the real projective space of dimension 3, $\mathbb{R}P(3)$ as follows: to any element $(v_1, v_2) \in G^o_{2,3}$ we associate a vector $\xi(v_1, v_2) \in \mathbb{R}^3$ as follows.

Denote by $A(v_1, v_2)$ the matrix whose columns one $v_1, v_2$ and the vectorial product $v_1 \wedge v_2 \in \mathbb{R}^3$. Then $A(v_1, v_2)$ is orthogonal and exhibits some eigenvector $\vec{u}(v_1, v_2)$ such that $\pm A(v_1, v_2) \cdot \vec{u}(v_1, v_2) = \vec{u}(v_1, v_2)$. Let $\pi(v_1, v_2)$ be the 2-dimensional subspace of $\mathbb{R}^3$ orthogonal to $\vec{u}(v_1, v_2)$

The restriction $A(v_1, v_2)|_{\pi(v_1, v_2)}$ is an orthogonal linear map of $\mathbb{R}^2$ so that it is a rotation of an angle say $\theta(v_1, v_2) \in [0, \pi]$.

If $\theta = 0$ we define $\xi(v_1, v_2) = 0 \in \mathbb{R}^3$ (in this case $A(v_1, v_2) = \text{Id}$). for $\theta(v_1, v_2) \in (0, \pi)$ we choose $\xi(v_1, v_2) \in \pi(v_1, v_2)$ with $|\xi(v_1, v_2)| = \theta$ and same direction and orientation that any $v \wedge A(v_1, v_2) \cdot v$ for $v \in \pi(v_1, v_2) - \{0\}$.

Finally, if $\theta(v_1, v_2) = \pi$. Then $v \wedge A(v_1, v_2) \cdot v = 0$, $\forall v \in \pi(v_1, v_2)$. In this case we cannot define an orientation for $\xi(v_1, v_2)$, what corresponds to identify the vectors $-v$ and $v \in \pi(v_1, v_2)$ having $|v| = \pi$.

Thus $\xi$ above define gives an homeomorphism $\xi \colon G^o_{2,3} \to \overline{B^3(0,\pi)}/ \sim$ of $G^o_{2,3}$ on to the quotient space $\overline{B^3(0,\pi)}/ \sim$ of the closed ball $\{(x_1, x_2, x_3) \in \mathbb{R}^3; \sum_{j=1}^{3} x_j^2 \leq \pi\}$ on $\mathbb{R}^3$ by the equivalence relation that identifies the points $v$ and $-v$ for $v \in \partial \overline{B^2(0,\pi)}$.

Clearly $\overline{B^3(0,\pi)}/ \sim$ is homeomorphic to $\mathbb{R}P(3)$ (recall that $\mathbb{R}P(3) \cong \mathbb{R}^3 \cup \mathbb{R}P(2)$).

The path $a = \tilde{a}(\,\cdot\,, 0) \colon \partial \overline{D}^2 \to G^o_{2,3}$ is according to this identification, a diameter of $\overline{B^3(0,\pi)}$ parallel to the $x_3$-axis, from down to up (orientation). Thus this path, once projected into $\mathbb{R}P(3)$, is not homotopic to a constant. This proves the theorem. □

**Remark 7.1.** The original proof of Lima is from 1963 and does not make use of Novikov's compact leaf theorem. Actually, the above proof shows:

**Theorem 7.2 (E. Lima, 1963).** *A compact simply-connected manifold of dimension three has rank one.*

**Remark 7.2.** The complete solution to the problem of describing the rank of closed 3-manifolds was given by Rosenberg-Roussarie [Rosenberg and Roussarie (1970)] where they prove that a rank two 3-manifold must be a non-trivial fiber bundle over the circle with a torus fiber.

**Exercise 7.0.1.** Is there any locally free action of the affine group Aff($\mathbb{R}$) on the 3-sphere?

# Chapter 8

# Tischler's theorem

## 8.1 Preliminaries

Let $M$ be a compact manifold admitting a submersion $f \colon M \xrightarrow{C^2} S^1$. We consider the angle-element 1-form $\theta \in H^1(S^1, \mathbb{R})$ and take $\omega = f^*(\theta)$ its lift to $M$. We obtain then a closed 1-form, without singularities, of class $C^1$ in $M$. Since $\omega$ is integrable, it defines a foliation $\mathcal{F}$ of codimension 1, class $C^1$ in $M$. Let now $p \in M$ be any point. Since $\omega$ is not singular there exist neighborhoods $p \in U_p \subset M$ and $C^1$ vector-fields $X_p$ in $U_p$ such that $\omega \cdot X_p = 1$ in $U_p$. Using partition of the unity we obtain finally a global vector-field $X$ in $M$ with the property that $\omega \cdot X = 1$. Since $M$ is compact, $X$ is complete defining therefore a flow $\varphi \colon \mathbb{R} \times M \to M$. From $\omega \cdot X = 1$ we conclude that the flow is transverse to $\mathcal{F}$. Since $d\omega = 0$ we have that $L_X(\omega) = d(\omega \cdot X) + i_X(d\omega) = 0$ so that $\varphi$ *preserves the foliation* $\mathcal{F}$ (each diffeomorphism $\varphi_t \colon M \to M$ takes leaves of $\mathcal{F}$ onto leaves of $\mathcal{F}$). We conclude that $\mathcal{F}$ is "invariant by a transverse flow". Tischler's theorem states the converse of this fact:

**Theorem 8.1 (Tischler-1970, [Tischler (1970)]).** *Let $M$ be a closed differentiable manifold. The following conditions are equivalent:*

(i) *$M$ supports a foliation $\mathcal{F}$, of class $C^1$ and codimension 1, invariant by a transverse flow $C^1$.*

(ii) *$M$ supports a closed 1-form of class $C^1$ without singularities.*

(iii) *$M$ fibers over the circle $S^1$.*

Taking into account the Theorem of Sacksteder (according to which a foliation of class $C^2$, codimension 1 and without holonomy is topologically conjugate to a foliation defined by a closed non-singular 1-form (cf. [Sack-steder (1965)]) we obtain in class $C^2$ the following equivalent condition:

(iv) *M admits codimension one foliation without holonomy.*

A demonstration of Tischler's theorem uses strongly the fact that in a closed manifold $M$ we can find closed differentiable 1-forms $\omega_1, \ldots, \omega_\ell \in H^1(M, \mathbb{R})$ such that given a base $\gamma_1, \ldots, \gamma_\ell$ of the free part of $H_1(M, \mathbb{Z})$ we have $\displaystyle\int_{\gamma_j} \omega_i = \delta_{ij}$ delta of Kronecker. Thus the closed 1-form closed $\omega$ in $M$ writes $\omega = \sum\limits_{j=1}^{\ell} \lambda_j \omega_j + df$ for some function $f \colon M \to \mathbb{R}$, where $\{\lambda_1, \ldots, \lambda_\ell\}$ generates the *group of periods* $\mathrm{Per}(\omega) < (\mathbb{R}, +)$ of $\omega$. If $\omega$ is non-singular then, since $\overline{\mathbb{Q}} = \mathbb{R}$, we can obtain perturbations $\omega' = \sum\limits_{j=1}^{\ell} \lambda_j \omega_j + df$ of $\omega$ such that $\omega'$ is non-singular and $\mathrm{Per}(\omega') \subset \mathbb{Q}$ and hence for some integral multiple $k \cdot \omega'$ we will $\mathrm{Per}(k\omega') \subset \mathbb{Z}$. Clearly $k\omega' = dg$ for some submersion $g \colon M \to \mathbb{R}/\mathbb{Z} = S^1$. $\blacksquare$

## 8.2   Proof of Tischler's theorem and generalizations

In this section we state the basic results we need in order to prove Tischler's theorem. Throughout this section $\mathcal{F}$ will denote a (non-singular) codimension one smooth foliation on a connected manifold $M$ of dimension $n \geq 2$.

**Definition 8.1.** Let $\varphi \colon \mathbb{C} \times M \to M$ be a smooth flow on $M$. We say that $\varphi$ is a flow *transverse* to $\mathcal{F}$ if the vector field $Z = \dfrac{\partial p}{\partial t}\bigg|_{t=0}$ (where $t \in \mathbb{C}$ is the complex time) is transverse to (the leaves of) $\mathcal{F}$.

We say that $\mathcal{F}$ is *invariant under* the flow $\varphi$ if each flow map $\varphi_t \colon M \to M$ takes leaves of $\mathcal{F}$ onto leaves of $\mathcal{F}$.

We shall say that $\mathcal{F}$ is *invariant under the transverse flow* of $Z$ if $Z$ is a complete vector field on $M$, whose corresponding flow $\varphi$ is transverse to $\mathcal{F}$ and $\mathcal{F}$ is invariant under $\varphi$.

**Example 8.1.** Let $M$ be a $n$-torus, $M = \mathbb{R}^n/\Lambda$ where $\Lambda \subset \mathbb{R}^n$ is some lattice. Let $\widetilde{\mathcal{F}}$ be the foliation on $\mathbb{R}^n$ by hyperplanes parallel to a given direction $\widetilde{Z} \in \mathbb{R}^n$. Then $\widetilde{\mathcal{F}}$ induces a foliation $\mathcal{F}$ on the quotient $M =$

$\mathbb{R}^n/\Lambda$ which is called a *linear foliation* on the Torus $M$. Such a foliation is invariant under a transverse flow given by a vector field $Z$ whose lift to $\mathbb{R}^n$ is $\widetilde{Z}$. As it is easily checked, $\mathcal{F}$ is given by a (non-singular) closed smooth 1-form $\Omega$ on $M$, with constant coefficients. The following (classic real) result states the existence of $\Omega$ as a general fact:

**Proposition 8.1.** *Let $\mathcal{F}$ be a smooth codimension one foliation invariant by a transverse smooth flow $\varphi$ of $Z$ on $M$. Then $\mathcal{F}$ is given by a (non-singular) closed smooth 1-form $\Omega$ characterized by:*

$$\int_{t_1}^{t_2} \Omega(\varphi_t(x)) \cdot Z(\varphi_t(x))\, dt = t_2 - t_1$$

$\forall x \in M, \ \forall t_1, t_2 \in \mathbb{R}.$

**Proof.** We follow the original construction in [Plante (1972)]. We construct $\Omega$ locally as a "time form" for $Z$. Given any point $p \in M$ choose a distinguished neighborhood $\xi \colon U \subset M \to \mathbb{R}^{n-1} \times \mathbb{R}$ such that $\xi$ takes $\mathcal{F}|_U$ into the horizontal foliation on $\mathbb{R}^{n-1} \times \mathbb{R}$. We may also assume that $\xi(p) = 0$ and (most important) $\xi(\varphi_t(p)) \in \mathbb{R}^{n-1} \times \{t\}$, $\forall t$ with $\varphi_t(p) \in U$ (here we use the fact that $\varphi$ is transverse to $\mathcal{F}$ and leaves $\mathcal{F}$ invariant). Define now $\Omega_U := d(\pi \circ \xi)$ where $\pi \colon \mathbb{R}^{n-1} \times \mathbb{R} \to \mathbb{R}$ is the projection $\pi(x, y) = y$. Given two such distinguished charts $\xi_j \colon U_j \subset M \to \mathbb{R}^{n-1} \times \mathbb{R}$ with $U_j$ connected and having connected intersection $U_1 \cap U_2 \neq \phi$ then if we put $\Omega_j := \Omega_{U_j} = d9\pi \circ \xi_j)$ we obtain in $U_1 \cap U_2$:

$$\Omega_1\big|_{U_1 \cap U_2} = d(\pi \circ \xi_1)\big|_{U_1 \cap U_2} \overset{(*)}{=} d(\pi \circ \xi_2)\big|_{U_1 \cap U_2}$$
$$= \Omega_2\big|_{U_1 \cap U_2}.$$

**Remark 8.1.** Notice that $\xi_j(\varphi_t(p_n)) \in \mathbb{R}^{n-1} \times \{t\}$ implies $\xi_j(\varphi_t(q)) \in \mathbb{R}^{n-1} \times \{t\}$, $\forall q \in L_{p_j} \cap U_j$ where $L_{p_j} = $ leaf of $\mathcal{F}$ through $p_j$, . Therefore $\xi_j(\varphi_t(q)) = (a_j(q, t), t)$, $\forall q \in L_{p_j} \cap U_j$ and henceforth $(\pi \circ \xi_j)(\varphi_t(q)) = \pi(a_j(q, t), t) = t$, $\forall q \in L_{p_j} \cap U_j$, $\forall t \approx 0$, so that finally
$$(\pi \circ \xi_j)(\varphi_t(r)) = t, \qquad \forall r \in U_j, \ \forall t \approx 0.$$

In this way we obtain a well-defined closed one-form $\Omega$ on $M$ which satisfies
$$\Omega(\varphi_t(p)) \cdot Z(\varphi_t(p)) = d(\pi \circ \xi)(\varphi_t(p)) \cdot Z(\varphi_t(p))$$
$$= \frac{d}{dt}\left((\pi \circ \xi)(\varphi_t(p))\right) =$$
$$= \frac{d}{dt}(t) = 1, \quad \forall p \in M, \ \forall t \in \mathbb{R}. \qquad \square$$

**Corollary 8.1.** *Let $\mathcal{F}$ be a codimension one (non-singular) smooth folia-tion on a compact (connected) manifold $M$. The following conditions are equivalent:*
*(i) $\mathcal{F}$ is invariant under some smooth transverse flow.*
*(ii) $\mathcal{F}$ is given by a closed smooth one-form $\Omega$ on $M$.*

**Proposition 8.2.** *Let $\mathcal{F}$, $\varphi$, $Z$, $\Omega$ be as in* Proposition 8.1 *but assume $M$ is compact. Given any leaf $L_0$ of $\mathcal{F}$ there exist a differentiable cover*

$$\sigma\colon L_0 \times \mathbb{R} \to M, \qquad \sigma(x,t) = \varphi_t(x);$$

*and an exact sequence of groups*

$$0 \longrightarrow \pi_1(L_0 \times \mathbb{R}) \xrightarrow{\sigma_\#} \pi_1(M) \longrightarrow A \longrightarrow 0,$$

*where $A$ is a finitely generated free abelian group. Moreover, $L_0$ is compact if, and only if, $A$ is a lattice on $\mathbb{R}$.*

**Proof.** Define

$$H = \left\{ [\gamma] \in \pi_1(M); \int_\gamma \Omega = 0 \right\}$$

then $H$ is a normal subgroup of $\pi_1(M)$ and it is free because

$$\int_{n\cdot\gamma} \Omega = n \cdot \int_\gamma \Omega \quad \forall \gamma \in \pi_1(M),\ \forall n \in \mathbb{Z}.$$

Put $A := \pi_1(M)/H$ then $A$ is finitely generated and also $A$ is abelian because $H \supset [\pi_1(M), \pi_1(M)]$ (the group of commutators) because

$$\int_{\gamma*\delta} \Omega = \int_\gamma \Omega + \int_\sigma \Omega = \int_\delta \Omega + \int_\gamma \Omega = \int_{\delta*\gamma}$$

$\forall \delta, \gamma \in \pi_1(M)$.
Let $P\colon \widetilde{M} \to M$ be the smooth cover of $M$, corresponding to $H$. Let also $\widetilde{\mathcal{F}}$, $\widetilde{\Omega}$ and $\widetilde{\varphi}_t$ be the lifting of $\mathcal{F}$, $\Omega$ and $\varphi_t$ to $\widetilde{M}$ respectively.

**Remark 8.2.** $\widetilde{\mathcal{F}} = P^*(\mathcal{F})$, $\widetilde{\Omega} = P^*(\Omega)$ are usual pull-backs. $\widetilde{\varphi}\colon \widetilde{M} \times \mathbb{R} \to \widetilde{M}$ is defined by

$$\widetilde{\varphi}_t(\tilde{p}) := \widetilde{\varphi_t(P(\tilde{p}))}, \quad \forall \tilde{p} \in \widetilde{M},\ \forall t \in \mathbb{R},$$

that is, for each $\tilde{\in}\widetilde{M}$, $\widetilde{\varphi}_t(\tilde{p})$ is the lifting by $P$ of the curve $\varphi_t(P(\tilde{p}))$ on $M$.

This lifting is well-defined because of the following:
Let $\gamma$, $\delta$ be simple piecewise smooth paths on $\mathbb{R}$ with $\gamma(0) = 0 = \delta(0)$ and $\delta(1) = t = \gamma(1)$.
Put $c = \delta^{-1} * \gamma$ then $c$ is closed. Since $\Omega$ is closed and $\varphi_p \colon \mathbb{R} \to M$ is smooth we have that $\int_c \varphi_p^*(\Omega) = 0$. Therefore,

$$\int_{(\varphi_p)_\# c} \Omega = 0, \quad \text{that is,} \quad \int_{\varphi_p(c)} \Omega = 0.$$

This says that $\varphi_p(c) \in H$. But $\varphi_p(c) = \varphi_p(\delta)^{-1} * \varphi_p(\gamma)$ so that $\varphi_p(\delta)$ and $\varphi_p(\gamma)$ are paths whose lifts by $P$ exhibit the same final points. Therefore we may define $\widetilde{\varphi} \colon \widetilde{M} \times \mathbb{R} \to \widetilde{M}$ in a natural way. It is now easy to check $\widetilde{\varphi}$ is (locally) smooth in each variable $\widetilde{x} \in \widetilde{M}$ and $t \in \mathbb{R}$ separately. By Hartogs' Theorem $\widetilde{\varphi}$ is smooth as a map $\widetilde{M} \times \mathbb{R} \to \widetilde{M}$. Finally we have by construction $P \circ \widetilde{\varphi}(t, \widetilde{x}) = \varphi_t(P(\widetilde{x}))$ so that

$$P \circ \widetilde{\varphi}(t, \widetilde{\varphi}(s, \widetilde{x})) = \varphi_t(P(\widetilde{\varphi}(s, \widetilde{x}))) = \varphi_t(\varphi_s(P(\widetilde{x})))$$
$$= \varphi_{t+s}(P(\widetilde{x})) = P \circ \widetilde{\varphi}(t + s, \widetilde{x}).$$

This implies that $\widetilde{\varphi}(t, \widetilde{\varphi}(s, \widetilde{x})) = \widetilde{\varphi}(t + s, \widetilde{x})$ so that $\widetilde{\varphi}$ is actually a flow on $\widetilde{M}$.
Let therefore $\widetilde{Z} = \left. \dfrac{\partial \widetilde{\varphi}}{\partial t} \right|_{t=0}$ be the corresponding smooth vector field. It is then clear that $P_* \widetilde{Z} = Z$, that is, $\widetilde{Z}$ is a lift of $Z$.
By construction if $\widetilde{\gamma} \in \pi_1(\widetilde{M})$ is such that $\int_{\widetilde{\gamma}} \widetilde{\Omega} = 0$ then $\widetilde{\gamma}$ is (homotopic to) the zero element so that $\widetilde{\Omega} = d\widetilde{f}$ for some smooth function $\widetilde{f} \colon \widetilde{M} \to \mathbb{R}$.

**Lemma 8.1.** *We have* $\widetilde{f}(\widetilde{\varphi}_t(\widetilde{x})) = t + \widetilde{f}(\widetilde{x}) \quad \forall t \in \mathbb{R}, \ \forall \widetilde{x} \in \widetilde{M}$.

**Proof.** Indeed,

$$\frac{d}{dt}\left(\widetilde{f}(\widetilde{\varphi}_t(\widetilde{x}))\right) = d\widetilde{f}(\widetilde{\varphi}_t(\widetilde{x})) \cdot \widetilde{Z}(\widetilde{\varphi}_t(\widetilde{x})) = \widetilde{\Omega}(\widetilde{\varphi}_t(\widetilde{x})) \cdot \widetilde{Z}(\widetilde{\varphi}_t(\widetilde{x})) = 1$$

and also

$$\widetilde{f}(\widetilde{\varphi}_t(\widetilde{x}))\big|_{t=0} = \widetilde{f}(\widetilde{x}). \qquad \square$$

Given any leaf $L_0$ of $\mathcal{F}$ on $M$ let $\widetilde{L}_0 \subset \widetilde{M}$ be a leaf of $\widetilde{\mathcal{F}}$ such that $P(\widetilde{L}_0) = L_0$. Define the map $g \colon \widetilde{L}_0 \times \mathbb{R} \to \widetilde{M}$ by setting $g(\widetilde{x}, t) = \widetilde{\varphi}_t(\widetilde{x})$.

**Lemma 8.2.** *$g$ is a smooth diffeomorphism of $\widetilde{L}_0 \times \mathbb{R}$ onto $\widetilde{M}$.*

**Proof.** We have

$$\frac{\partial g}{\partial t}(\tilde{x}_0, t_0) = \left.\frac{\partial \tilde{\varphi}_t}{\partial t}\right|_{t=t_0}(\tilde{x}_0) = \tilde{Z}(\tilde{\varphi}_{t_0}(\tilde{x}_0)).$$

Also

$$\frac{\partial g}{\partial \tilde{x}}(\tilde{x}_0, t_0) = \left.\frac{\partial}{\partial \tilde{x}}(\tilde{\varphi}_{t_0}(\tilde{x}))\right|_{\tilde{x}=\tilde{x}_0} = \frac{\partial \tilde{\varphi}}{\partial \tilde{x}}(\tilde{x}_0, t_0).$$

Since the flow of $Z$ is transverse to $\mathcal{F}$ it follows that the flow $\tilde{\varphi}$ is transverse to $\tilde{\mathcal{F}}$ so that $g$ is a local diffeomorphism in $\tilde{L}_0 \times \mathbb{R}$.

Now we notice that if $g(\tilde{x}_1, t_1) = g(\tilde{x}_2, t_2)$ then $\tilde{\varphi}_{t_1}(\tilde{x}_1) = \tilde{\varphi}_{t_2}(\tilde{x}_2)$ and $\tilde{f}(\tilde{\varphi}(t_1, \tilde{x}_1)) = \tilde{f}(\tilde{\varphi}(t_2, \tilde{x}_2))$ so that $\tilde{\varphi}_{t_1-t_2}(\tilde{x}_1) = \tilde{x}_2$ and $t_1 + \tilde{f}(\tilde{x}_1) = t_2 + \tilde{f}(\tilde{x}_2)$.

Now, $x_1$ and $\tilde{x}_2$ belong to the same leaf $\tilde{L}_{\tilde{x}_0} = \tilde{L}_0$ of $\tilde{\mathcal{F}}$ so that $f(\tilde{x}_1) = f(\tilde{x}_2)$, this implies $t_1 = t_2$ and therefore $\tilde{x}_1 = \tilde{x}_2$.

Therefore $g$ is also injective and it is a diffeomorphism of $\tilde{L}_0 \times \mathbb{R}$ onto its image $g(\tilde{L}_0 \times \mathbb{R}) \subset \widetilde{M}$. It remains to prove that $g(\tilde{L}_0 \times \mathbb{R}) = \widetilde{M}$.

It is enough to prove that this image of $g$ is closed. Take any point $\tilde{x}_1 \in \widetilde{M}$ belonging to the closure of $g(\tilde{L}_0 \times \mathbb{R})$ in $\widetilde{M}$. Let $\tilde{B}_1 \ni \tilde{x}_1$ be any open ball in the leaf $\tilde{L}_1 \ni \tilde{x}_1$. Let $\tilde{U}$ be the "cylinder" $\tilde{U} = \bigcup_{t\in\mathbb{R}} \tilde{\varphi}_t(\tilde{B}_1)$, and take any

$$\tilde{x} \in \tilde{U} \cap g(\tilde{L}_0 \times \mathbb{R}).$$

We have $\tilde{x} \in \tilde{\varphi}_s(\tilde{L}_0)$ for some $s \in \mathbb{R}$ and also there exists $r \in \mathbb{R}$ such that $\tilde{x} \in \tilde{\varphi}_r(\tilde{B}_1)$. Thus $\tilde{x}_1 \in \tilde{\varphi}_{s-r}(\tilde{L}_0)$ and hence $\tilde{x}_1 \in g(\tilde{L}_0 \times \mathbb{R})$. This proves the lemma. $\qquad\square$

Now we may prove:

**Lemma 8.3.** $P|_{\tilde{L}_0} \colon \tilde{L}_0 \subset \widetilde{M} \to L_0 \subset M$ *is a bijection and therefore a diffeomorphism.*

**Proof.** The map $P$ is injective, for if $\tilde{x}_1, \tilde{x}_2 \in \tilde{L}_0$ are such that $P(\tilde{x}_1) = P(\tilde{x}_2)$ then we may take a path $\tilde{\alpha} \colon [0, 1] \to L_0$ of class $C^1$ with $\tilde{\alpha}(0) = \tilde{x}_1$ and $\tilde{\alpha}(1) = \tilde{x}_2$.

This gives a projected path $\alpha = P \circ \tilde{\alpha} \colon [0, 1] \to L_0 \subset M$ which is closed, i.e., $\alpha \in \pi_1(L_0)$.

We have

$$\frac{d}{dt}(\tilde{f}(\tilde{\alpha}(t))) = \tilde{\Omega}(\tilde{\alpha}(t)) \cdot \tilde{\alpha}'(t) \quad \text{so that}$$

$$\tilde{\Omega}(\tilde{\alpha}(t)) \cdot \tilde{\alpha}'(t) = 0, \quad \forall t \in [0, 1] \quad \text{and therefore}$$

$$\Omega(\alpha(t)) \cdot \alpha'(t) = 0, \quad \forall t \in [0,1].$$

This gives

$$0 = \int_0^1 \Omega(\alpha(t)) \cdot \alpha'(t) \, dt = \int_\alpha \Omega$$

and therefore $[\alpha] \in H \subset \pi_1(M)$. This gives $\tilde{\alpha}(0) = \tilde{\alpha}(1)$ in $\widetilde{M}$, i.e., $\tilde{x}_1 = \tilde{x}_2$. $\qquad \square$

Let now $\eta_t \colon L_0 \times \mathbb{R} \to L \times \mathbb{R}$ be given by $\eta_t(x,s) := (x, s+t)$. Let also $G \colon L_0 \times \mathbb{R} \to \widetilde{M}$ be defined by

$$G(x,s) := g\big((P|_{\tilde{L}_0})^{-1}(x), s\big).$$

Notice that

$$\big((P|_{\tilde{L}_0})^{-1}(x), s\big) \in \tilde{L}_0 \times \mathbb{R}.$$

Consider the following diagram

$$
\begin{array}{ccccc}
M & \xleftarrow{\ P\ } & \widetilde{M} & \xleftarrow{\ G\ } & L_0 \times \mathbb{R} \\
\varphi_t \downarrow & & \tilde{\varphi}_t \downarrow & & \downarrow \eta_t \\
M & \xleftarrow{\ P\ } & \widetilde{M} & \xleftarrow{\ G\ } & L_0 \times \mathbb{R}
\end{array}
$$

The left side is commutative by construction.
Now we observe that given $(x,s) \in L_0 \times \mathbb{R}$ we have

$$
\begin{aligned}
G(\eta_t(x,s)) = G(x, s+t) &= g\big((P|_{\tilde{L}_0})^{-1}(x), s+t\big) \\
&= \tilde{\varphi}_{s+t}\big((P|_{\tilde{L}_0})^{-1}(x)\big) \\
&= \tilde{\varphi}_t\big(\tilde{\varphi}_s\big((P|_{\tilde{L}_0})^{-1}(x)\big)\big)
\end{aligned}
$$

$$\Rightarrow \ G(M_t(x,s)) = \tilde{\varphi}_t(G(x,s)) = (\tilde{\varphi}_t \circ G)(x,s).$$

Therefore the whole diagram is commutative.
Define now

$$\sigma := P \circ G \colon L_0 \times \mathbb{R} \to M$$

by requiring tht the diagram below is commutative:

$$
\begin{array}{ccc}
M & \xleftarrow{\ \sigma\ } & L_0 \times \mathbb{R} \\
\varphi_t \downarrow & & \downarrow \eta_t \\
M & \xleftarrow{\ \sigma\ } & L_0 \times \mathbb{R}
\end{array}
$$

In other words:

$$\varphi_t \circ \sigma(x,s) = \sigma(x, s+t).$$

$$\square$$

**Lemma 8.4.** $\sigma\colon L_0 \times \mathbb{R} \to M$ is a covering map.

**Proof.** We know that $P$ is a covering map $P\colon \widetilde{M} \to M$. It is therefore enough to show that $G\colon L_0 \times \mathbb{R} \to \widetilde{M}$ is a covering map. Actually $g\colon \widetilde{L}_0 \times \mathbb{R} \to \widetilde{M}$ is a diffeomorphism and so is $P_{\widetilde{L}_0}\colon \widetilde{L}_0 \to L_0$, so that $G\colon L_0 \times \mathbb{R} \to \widetilde{M}$ is a diffeomorphism. $\square$

Clearly $\sigma(x, t) = \varphi_t \circ \sigma(x, 0) = \varphi_t(x)$, $\forall t \in \mathbb{R}$ $\forall x \in L_0$. Therefore $\sigma$ satisfies the first condition in the statement of Proposition 8.3.
If for any $[\gamma] \in \pi_1(M)$ we have $[\gamma] = \sigma_\#([\alpha])$ in $\pi_1(M)$, for some $[\alpha] \in \pi_1(L_0 \times \mathbb{R})$ then

$$\int_\gamma \Omega = \int_{\sigma \circ \alpha} \Omega = \int_{\alpha \subset L_0 \times \mathbb{R}} \sigma^*(\Omega) = \int_\alpha (\varphi_t)^*(\Omega) \underset{\substack{\text{for } \alpha \subset L_0 \times \mathbb{R} \\ \text{and } \Omega = 0 \text{ along } L_0}}{=} 0$$

so that $[\gamma] \in H$. Conversely, if $[\gamma] \in H$ then $\gamma = P_\#(\widetilde{\gamma})$ for some $\widetilde{\gamma} \in \pi_1(\widetilde{M})$ and therefore we have $\widetilde{\gamma} = g_\#(\widetilde{\alpha})$ for some $\widetilde{\alpha} \in \pi_1(\widetilde{L}_0 \times \mathbb{R})$ so that $\gamma = P_\#(g_\#(\widetilde{\alpha})) \Rightarrow \gamma = (P \circ G)_\#(\alpha)$ where $\alpha = P \circ \widetilde{\alpha} \in \pi_1(L_0 \times \mathbb{R})$ is obtained in a natural way.
Therefore we have proved the following:

**Lemma 8.5.** *The sequence below is exact*

$$0 \longrightarrow \pi_1(L_0 \times \mathbb{R}) \xrightarrow{\sigma_\#} \pi_1(M) \longrightarrow A \longrightarrow 0.$$

**Remark 8.3.** Another way of seeing the above equivalence is the following: if $[\gamma] \in H$ then $\gamma \in \pi_1(M)$ is such that $\int_\gamma \Omega = 0$. Therefore we may consider the lifting $\alpha$ of $\gamma$ by $\sigma$ to $L_0 \times \mathbb{R}$ obtaining a path such that $\int_\alpha (\varphi_t)^*(\Omega) = 0$ and therefore $\alpha$ is closed that is, $[\gamma] = \sigma_\#([\alpha])$ for $[\alpha] \in \pi_1(L_0 \times \mathbb{R})$.

Assume that $A$ has rank one, $A \approx \mathbb{Z}$. We may take a transformation $T\colon L_0 \times \mathbb{R} \to L_0 \times \mathbb{R}$ which corresponds to a generator of $A = \pi_1(M)/H$ (notice that the covering $\sigma\colon L_0 \times \mathbb{R} \to M$ has group isomorphic to $\pi_1(M)/H$ because of the exact sequence $(0 \to \pi_1(L_0 \times \mathbb{R}) \xrightarrow{\sigma_\#} \pi_1(M) \to \pi_1(M)/H \to 0)$.

$$L_0 \times \mathbb{R} \xrightarrow{T} L_0 \times \mathbb{R}$$

The diagram $\quad \sigma \searrow \qquad \swarrow \sigma \quad$ commutes.
$$M$$

Let $T_0 := T\big|_{L_0 \times \{0\}} : L_0 \times \{0\} \to L_0 \times \mathbb{R}$.

**Lemma 8.6.** $\exists t_0 \in \mathbb{R} - \{0\}$ such that $T(L_0 \times \{0\}) = L_0 \times \{t_0\}$.

**Proof.** Write $T(x,t) = (a(x,t), b(x,t))$ so that $\sigma \circ T = \sigma \Rightarrow$ $\varphi_{b(x,t)}(a(x,t)) = \varphi_t(x) \Rightarrow \varphi_{b(x,0)}(a(x,0)) = 0 \ \forall x \in L_0.$ (*)
We have $a(x,0) \in L_0$, $\forall x \in L_0$ therefore (since $\varphi_t$ is transverse to $\mathcal{F}$) we must have from (*) from ($b(x,0)$ is constant $\forall x \in L_0$ and therefore if we put $t_0 = b(x,0)$ then

$$\varphi_{t_0}(a(x,0)) = x, \quad \forall x \in L_0.$$

That by

$$T(L_0 \times \{0\}) \subset L_0 \times \{t_0\}. \quad \square$$

If $t_0 = 0$ then $b(x,0) = 0$, $\forall x \in L_0$ and $T(x,0) = (a(x,0),0)$ and also (*) $a(x,0) = x \ \forall x \in L_0$.
Thus $T(x,0) = (x,0)$, $\forall x \in L_0$. This is not possible for $T$ is a non trivial covering transformation. $\qquad \square$

Define now a map $f: M \to \mathbb{R}/t_0\mathbb{Z} \simeq S^1$ by setting $f(x) := s$ (mod $t_0$) where $x \in \varphi_s(L_0)$. Notice that given any $x_1 \in M$, since $\sigma: L_0 \times \mathbb{R} \to M$, $\sigma(x,t) = \varphi_t(x)$ is a covering, it follows that $x_1 \in \varphi_s(L_0)$ for some $s \in \mathbb{R}$.
Now, if $s_1, s_2 \in \mathbb{R}$ are such that $x \in \varphi_{s_j}(L_0)$, $j = 1,2$, then $\exists x_1, x_2 \in L_0$ with $x = \varphi_{s_1}(x_1)$, $x = \varphi_{s_2}(x_2)$ so that $\sigma(x_1, s_1) = x = \sigma(x_2, s_2)$. Since the group of covering maps of $\sigma$ is generated by $T$ we must have $(x_2, s_2) = T^n(x_1, s_1)$ for some $n \in \mathbb{Z}$ so that $s_2 = s_1 + n \cdot t_0$, so that $s_2 = s_1$ (mod $t_0$). Therefore $f: M \to S^1 = \mathbb{R}/t_0\mathbb{Z}$ is well-defined and clearly smooth.
Since $\varphi_t$ takes leaves of $\mathcal{F}$ onto leaves, $f$ is constant along the leaves of $\mathcal{F}$. Thus $f$ is a smooth first integral for $\mathcal{F}$.
Assume now that rank$(A) = 0$. In this case $H = \pi_1(M)$ and $P: \widetilde{M} \to M$ is the universal covering of $M$. Since $A = \{0\}$ we have a diffeomorphism $M \overset{\sigma}{\simeq} L_0 \times \mathbb{R}$ which is not possible because $M$ is compact.
Conversely, assume now that $L_0$ is a compact leaf of $\mathcal{F}$. Since $\mathcal{F}$ has trivial holonomy the Stability Theorem of Reeb (see [Godbillon (1991)]) implies that all the leaves of $\mathcal{F}$ are compact and $\mathcal{F}$ is a Seifert (smooth) fibration. Now, the group $A$ acts on $L_0 \times \mathbb{R}$ taking leaves of $\sigma^*(\mathcal{F})$ onto leaves of $\sigma^*(\mathcal{F})$ in a natural way as in Lemma 8.5. Therefore, since $\mathcal{F}$ is a compact foliation, the leaves of $\sigma^*(\gamma)$ are closed on $L_0 \times \mathbb{R}$ and therefore the action of $A$ must

be discrete so that indeed, $A$ must correspond to a discrete subgroup of $\mathbb{R}$ and therefore rank$(A) \leq 1$ as it is well-known (see Remark 8.4 below).

**Remark 8.4.** We remark that, according to what we have seen, we have an homomorphism of groups

$$\xi \colon \pi_1(M) \to (\mathbb{R}, +)$$

$$[\gamma] \mapsto \int_\gamma \Omega$$

whose kernel is $H$ so that there exists an injective homomorphism $\bar{\xi} \colon A = \pi_1(M)/H \to \mathbb{R}$, so that $A$ is naturally identified to a certain subgroup of $(\mathbb{R}^2, +)$.

Therefore we have proved that rank$(A) = 1 \Leftrightarrow L_0$ is compact $\Leftrightarrow$ all leaves of $\mathcal{F}$ are compact. This ends the proof of Proposition 8.2.     □

**Corollary 8.2.** *Let $M$ be compact, $\mathcal{F}$, $\varphi_t$, $A$ as in* Proposition 8.2. *Then $1 \leq$ rank$(A) \leq$ rank$(H_1(M, \mathbb{Z}))$. Moreover, if $M$ is an orientable compact manifold and* rank$(H_1(M, \mathbb{R})) \leq 1$ *then $\mathcal{F}$ is a foliation by compact leaves.*

Next step is the following:

**Proposition 8.3.** *Let $\mathcal{F}$, $\varphi_t$, $\Omega$, $A$ and $M$ be as in* Proposition 8.2. *Let* Per$(L_0) := \{t \in \mathbb{R}; \varphi_t(L_0) = L_0\}$ *and $\eta_t \colon L_0 \times \mathbb{R} \to L_0 \times \mathbb{R}$ be given by $\eta_t(x, s) = (x, s + t)$.*
*(i) If $T \colon L \times \mathbb{R} \to L_0 \times \mathbb{R}$ is a covering transformation of the covering*

$$\sigma \colon L_0 \times \mathbb{R} \to M$$

$$(x, t) \mapsto \varphi_t(x)$$

*then $T(L_0 \times \{t\}) = \eta_{t_0}(L_0 \times \{0\})$ for some $t_0 = t_0(T) \in \mathbb{R}$.*
*(ii) The correspondence $T \mapsto t_0(T)$ defines an isomorphism $A \to$ Per$(L_0)$.*
*(iii)* Per$(L_0)$ *is the group of periods of $\Omega$.*

**Proof.** As we have seen in the proof of Proposition 8.2 above for each covering transformation $T$ of $\sigma$ we must have $T(L_0 \times \{0\}) = L_0 \times \{t_0(T)\}$ for some $t_0(T) \in \mathbb{R}$. Moreover $t_0(T) = 0$ if, and only if, $T$ is the identity. It is also possible to see that $t_0(T)$ depends only on $T$, not on the choice of the leaf $L_0 \subset M$. Therefore we have $T(L \times \{0\}) = \eta_{t_0(T)}(L \times \{0\})$ for any leaf $L$ of $\mathcal{F}$.

The mapping $\xi \colon A \to \mathbb{R}$, $T \mapsto t_0(T)$ is therefore such that $\xi(A) \subset \mathrm{Per}(L_0)$: given any $x_0 \in L_0$ we have $T(L_0 \times \{0\}) = L_0 \times \{t_0(T)\} \Rightarrow T(x_0, 0) \in L_0 \times \{t_0(T)\} \Rightarrow$ if we write $T(x, t) = (a(x, t), b(x, t))$ then $T(x, 0) = (a(x, 0), t_0(T))$ and therefore

$$x_0 = \varphi_0(x_0) = \sigma(x_0, 0) = \sigma \circ T(x_0, 0)$$

$$\sigma(a(x_0, 0), t_0(T)) = \varphi_{t_0(T)}(a(x_0, 0))$$

so that $\varphi_{t_0(T)}(L_0) = L_0$ and then $t_0(T) \in \mathrm{Per}(L_0)$. Thus we have $\xi \colon A \to \mathrm{Per}(L_0) \subset \mathbb{R}$.

**Lemma 8.7.** $\xi$ *is an injective group homomorphism.*

**Proof.** Given $S, T \in A$ be have $\xi(S \circ T) = t_0(S \circ T)$ and by definition

$$S \circ T(L_0 \times \{0\}) = L_0 \times \{t_0(S \circ T)\}.$$

But, on the other hand,

$$S \circ T(L_0 \times \{0\}) = S(T(L_0 \times \{0\})) = S(L_0 \times \{t_0(T)\})$$

$$\Rightarrow S \circ T(L_0 \times \{0\}) = S(L_0 \times \{t_0(T)\}).$$

Now, for any leaf $L$ of $\mathcal{F}$ we have $S(L \times \{0\}) = L \times \{t_0(S)\}$. Therefore

$$S \circ T(L_0 \times \{0\}) = L_0 \times \{t_0(S) + t_0(T)\}.$$

This implies that $t_0(S \circ T) = t_0(S) + t_0(T)$. The injectivity of $\xi$ we have already checked. $\qquad\square$

Finally we claim that $\xi$ is surjective. Indeed, given any $t_0 \in \mathrm{Per}(L_0)$ and any $x_0 \in L_0$ we may consider paths $\alpha := \varphi_{st_0}(x_0)$ in $M$ and $\beta$ in $L_0$, joining $l_0 \ni \varphi_{t_0}(x_0)$ to $x_0$ because $t_0 \in \mathrm{Per}(L_0)$.

The homotopy class

$$[\gamma] = [\alpha * \beta] \in \pi_1(M)$$

is such that if $T \in A$ corresponds to $[\gamma]$ then $T(L_0 \times \{0\}) = L_0 \times \{t_0(T)\}$ where $t_0(T)$ is given by

$$t_0(T) = \int_\gamma \Omega = \int_\alpha \Omega + \int_\beta \Omega = \int_\alpha \Omega$$

$$= \int_0^1 \Omega(\varphi_{st_0}(x_0)) \cdot \frac{d}{ds}(\varphi_{st_0}(x_0))\, dx = t_0 \int_0^1 1 \cdot ds = t_0.$$

Thus $t_0(T) = t_0$ and $\xi$ is surjective. This shows (i) and (ii). $\qquad\square$

**Lemma 8.8.** Per($L_0$) *is the group of periods of* $\Omega$ *which is defined by*

$$\mathrm{Per}(\Omega) := \left\{ \int_\gamma \Omega; [\gamma] \in \pi_1(M) \right\} < \mathbb{R}.$$

**Proof.** Let $t_0 \in \mathrm{Per}(L_0)$ and $[\gamma] = [\alpha * \beta]$ as above, then $\int_{[\gamma]} \Omega = t_0$ so that $t_0 \in \mathrm{Per}(\Omega)$. Conversely, given any period $t_0 \in \mathrm{Per}(\Omega)$ say $t_0 = \int_\gamma \Omega$ for some $[\gamma] \in \pi_1(M)$ we may perform small homotopies so that $[\gamma]$ is of the form $[\gamma] = [\alpha_1 * \beta_1 * \cdots * \alpha_r * \beta_r]$ with $\alpha_j$ segment of orbit of $\varphi_t$ and $\beta_j$ contained in a single leaf of $\mathcal{F}$, $\forall j \in \{1, \ldots, r\}$. Using the flow we may obtain a homotopy between $\beta_{r-1} * \alpha_r$ and some path of the form $\alpha * \beta$. Using the flow: $\beta_{r-1} * \alpha_r$ is homotopic to some path of the form $\alpha * \beta$. Therefore we may assume that $r = 1$, and $\gamma = \alpha_1 * \beta_1$. Therefore

$$t_0 = \int_\gamma \Omega = \int_{\alpha_1} \Omega \Rightarrow \alpha_1(t_0) = \varphi_{t_0}(x_0)$$

and $\alpha_1(0) = x_0$ belong $t_0$ a same leaf of $\mathcal{F}$ and therefore $t_0 \in \mathrm{Per}(L_0)$. This proves (iii) and Proposition 8.3. $\quad\square$

**Corollary 8.3.** *Let* $\mathcal{F}$, $\Omega$, $\varphi_t$, $A$, $\mathrm{Per}(\Omega)$, $M$ *compact be as above. The leaves of* $\mathcal{F}$ *are compact if, and only if,* $\mathrm{Per}(\Omega) \subset \mathbb{R}$ *has rank one and defines a lattice on* $\mathbb{R}$. *In any other case the leaves of* $\mathcal{F}$ *are not closed.*

**Proof.** We have already proved that $\mathrm{rank}(A) \geq 1$ and also $\mathrm{rank}(A) = 1$ and if, and only if, $\mathcal{F}$ is a compact foliation. Moreover $\mathrm{rank}(A) \geq 2$ implies $\mathrm{Per}(\Omega) \subset \mathbb{R}$ is not discrete in fact it is dense, what implies (see Remark 8.5) that $A$ acts in the leaves of $\sigma^*(\mathcal{F})$ in $L_0 \times \mathbb{R}$ with non-discrete dynamics. This implies that the leaves of $\mathcal{F}$ are not closed. $\quad\square$

**Remark 8.5.** Let $M$ be a compact differentiable manifold supporting a non-singular codimension one smooth foliation invariant by a transverse flow. Then $\pi_1(M)$ is not finite, indeed $\mathrm{rank}(H_1(M, \mathbb{Z})) \geq 1$.

Indeed, if $\pi_1(M)$ is finite then the universal covering $\widetilde{M}$ of $M$ is also compact so that the closed one-form $\Omega$ lifts into a closed non-zero smooth 1-form $\widetilde{\Omega}$ on $\widetilde{M}$ which is exact, $\widetilde{\Omega} = d\tilde{f}$ for some smooth function $\tilde{f} \colon \widetilde{M} \to \mathbb{R}$. Since $\widetilde{M}$ is compact $\tilde{f}$ must exhibit some critical point is constant and $\Omega$ has some singularity, contradiction. $\quad\square$

Now we are in conditions to prove Tischler's theorem.

**Proof of Tischler's theorem.** We may assume that $M$ is orientable and oriented. According to what we have seen above the foliation $\mathcal{F}$ is given by a non-singular smooth closed one-form $\Omega$ in $M$. We may find a basis $\{w_j = \mu^*(\alpha_j)\}$ of the group the De Rham cohomology group $H^1(M, \Omega^1)$ given by (classes of) closed 1-forms in $M$ such that

for some loops $\gamma_1, \ldots, \gamma_r$ corresponding to a basis of the free part of $H^1(M, \mathbb{Z})$ we have

$$\int_\gamma w_i = \delta_{ij} \, .$$

We may therefore write $\Omega = \sum\limits_{j=1}^{r} \lambda_j w_j + df$ for some $\lambda_j \in \mathbb{R}$, and some $f \colon M \to \mathbb{R}$ smooth. Then $\lambda_j = \int_{\gamma_j} \Omega$ so that $\{\lambda_1, \ldots, \lambda_r\} \subseteq \mathrm{Per}(\Omega)$. Indeed $\mathrm{Per}(\Omega)$ is generated (as a group) by the $\lambda_j$'s, $j = 1, \ldots, r$, *i.e.*, $\mathrm{Per}(\Omega) = \langle \{\lambda_1, \ldots, \lambda_r\} \rangle$. Let now $(\lambda'_1, \ldots, \lambda'_r) \in \mathbb{R}^r$ be such that $\Omega' := \sum\limits_{j=1}^{r} \lambda'_j \cdot w_j$ is close enough to $\Omega$ so that it is also non singular (recall that $\Omega$ is non singular and $M$ is compact) and the subgroup $\langle \{\lambda'_1, \ldots, \lambda'_r\} \rangle$ of $\mathbb{R}$ is a rank 1 discrete lattice (it is enough to choose $\{\lambda'_1, \ldots, \lambda'_r\} \subseteq \mathbb{Q}$ of rank 1). Thus $\Omega'$ defines a fibration of $M$ over the circle $S^1 \cong \mathbb{R}/\Lambda'$, $\Lambda' = \langle \{\lambda'_1, \ldots, \lambda'_r\} \rangle$. $\quad\square$

# Chapter 9

# Plante's compact leaf theorem

This chapter is dedicated to an exposition of the celebrated result by J. Plante [Plante (1975)], asserting that every codimension one foliation on a closed manifold $M$ exhibiting a leaf of subexponential growth also exhibits a leaf of polynomial growth of degree $\leq \max\{0, \beta_1(M) - 1\}$, where $\beta_1(M)$ is the first Betti number of $M$. In particular, *the foliation has a compact leaf provided that it exhibits some leaf of subexponential growth on compact $M$ having $\beta_1(M) \leq 1$.*

## 9.1 Growth of foliations and existence of compact leaves

In opposition to a $n$-dimensional version of Novikov's compact leaf theorem there are examples of foliations $C^2$ of codimension one of $S^n$, $n \geq 4$ without compact leaves. In particular, the minimal set of such a foliation is exceptional (cf. [Sacksteder (1965)]). It was J. Plante, in an outstanding work, who initiated the modern comprehension of such facts relating the concepts of growth of leaves and existence of exceptional minimal sets (cf. [Plante (1973)]). Let us recall such concepts:

### 9.1.1 *Growth of Riemannian manifolds*

Let $(M, g)$ be a connected oriented Riemannian manifold of class $C^r, r \geq 1$. Given any point $x \in M$ the *growth function* of $M$ at $x$ is defined by $\gamma_x(r) :=$ volume of the closed metric ball $B[x; r]$. The growth type of $\gamma_x$ does not depend on the choice of $x \in M$. In this way we may introduce of *polynomial growth, exponential growth, ...* for $(M, g)$. If $M$ is compact then it has polynomial growth of degree zero. In the case $r = \infty$ we have the following:

**Proposition 9.1 (Moussu, Pelletier [Godbillon (1991)]).** *For any $r \geq 0$ and any $x \in M$ the closed ball $B[x;r]$ is a standard Whitney domain: the boundary $\partial B[x;r]$ contains a compact subset $K$ with zero $(m-1)$-dimensional measure such that $\partial B[x;r] - K$ is a submanifold with boundary of $M$. Moreover, the function $\gamma_x(r) = \mathrm{vol}(B[x;r])$ is differentiable with respect to $r$ and its derivative at the point $r_o$ is the volume of the $(m-1)$-dimensional sphere $\partial B[x;r_o]$.*

It follows from the above result that *if* $\lim\limits_{r \to +\infty} \inf \frac{\mathrm{vol}(\partial B[x;r])}{\mathrm{vol}(B[x;r])} > 0$ *then* $M$ *has exponential growth.*

### 9.1.2   Growth of leaves

The notion of growth for the leaves of a foliation on a *compact* manifold may be introduced in a geometric way regarding the growth of the volume of the balls in the leaves, and it will be related to the growth of the orbits of the holonomy pseudogroup of the foliation, as we will see. The main remark is the following:

**Proposition 9.2 ([Godbillon (1991)]).** *Given two Riemannian metrics in a compact manifold $M$ equipped with a $C^1$ regular foliation $\mathcal{F}$, the metrics induce on each leaf $L$ of $\mathcal{F}$, complete quasi-isometric metrics. Therefore, the growth type of the leaf $L$ does not depend on the choice of the ambient metric.*

In the non-compact case however, we may fix the metric and consider the growth type of the leaves with respect to this fixed metric. Let $(M, g)$ be a Riemannian manifold, perhaps non-compact, and let $\mathcal{F}$ be a (regular) $C^1$ foliation of codimension $k$ on $M$. Assume that $M$ is oriented and $\mathcal{F}$ is transversely oriented. For each $x \in M$ denote by $L_x$ the leaf of $\mathcal{F}$ through $x$. The metric on $M$ induces a metric $g_x$ along the (immersed) leaf $L_x$.

**Definition 9.1.** The *growth type* of the leaf $L_x$ with respect to the metric $g$ is the growth type of the Riemannian manifold $(L_x, g_x)$.

Therefore, compact leaves have polynomial growth of degree zero.

### 9.1.3   Growth of orbits

Let $X$ be a Hausdorff topological space and $\Gamma$ a collection of homeomorphisms $g : U \to V$, where $U, V$ are open subsets of $X$. Denote by $\mathrm{Dom}(g)$

and Range($g$) the domain and the range of $g \in \Gamma$ respectively.

**Definition 9.2 ([Plante (1972)]).** $\Gamma$ is a *pseudo-group* of local homeomorphisms of $X$ if:

(i) For any $g \in \Gamma$ we have $g^{-1} \in \Gamma$ and $\mathrm{Dom}(g) = \mathrm{Range}(g^{-1})$, and $\mathrm{Dom}(g^{-1}) = \mathrm{Range}(g)$.

(ii) If $g_1, g_2 \in \Gamma$ and $g \colon \mathrm{Dom}(g_1) \cup \mathrm{Dom}(g_2) \to \mathrm{Range}(g_1) \cup \mathrm{Range}(g_2)$ is a homeomorphism such that $g\big|_{\mathrm{Dom}(g_i)} = g_i, i = 1, 2$, then $g \in \Gamma$.

(iii) $Id \colon X \to X$ belongs to $\Gamma$.

(iv) If $g_1, g_2 \in \Gamma$ then $g_1 \circ g_2 \in \Gamma$, with $\mathrm{Dom}(g_1 \circ g_2) \subset g_1^{-1}(R(g_2)) \cap \mathrm{Dom}(g_1)$.

(v) If $g \in \Gamma$ and $U \subset \mathrm{Dom}(g)$ is an open subset then $g\big|_U \in \Gamma$.

The *orbit* of $x$ in the pseudogroup $\Gamma$ is defined by $\Gamma(x) := \{g(x) \in X, g \in \Gamma, x \in \mathrm{Dom}(g)\}$. Assume now that $\Gamma$ is finitely generated by a (symmetric) finite subset $\Gamma^o \subset \Gamma$.

**Definition 9.3.** For $x \in X$ and $n \in \mathbb{N}$ we define $\Gamma_n(x) := \{y \in X, y = g_{\alpha_1} \circ \ldots \circ g_{\alpha_k}(x), k \leq n, g_{\alpha_j} \in \Gamma^o, j = 1, \ldots, k\}$. The *growth type* of the orbit of $x$ in $\Gamma$ is the growth type of the function $\gamma_x(n) := \sharp\Gamma_n(x)$ as $n \in \mathbb{N}$.

### 9.1.4 Combinatorial growth of leaves

Let $M$ be a compact manifold and $\mathcal{F}$ a foliation of codimension $k$ on $M$. Given a finite cover $\mathcal{U} = \{U_1, \ldots, U_r\}$ of $M$ by distinguished neighborhoods for $\mathcal{F}$ we denote by $\Gamma_{\mathcal{U}}$ the holonomy pseudogroup associated to this cover. Then $\Gamma_{\mathcal{U}}$ is finitely generated. Given a point $x \in U_1$ and $n \in \mathbb{N}$ the value of the growth function $\gamma_{\mathcal{U},x}(n)$ is equal to the number of plaques of $\mathcal{U}$ that can be joined to the plaque $P_{1,x}$ of $U_1$, by a chain of plaques with at most $n$ plaques.

**Proposition 9.3 ([Godbillon (1991)]).** *Let $\mathcal{U}$ and $\mathcal{V}$ be finite covers by distinguished neighborhoods of the manifold $M$. The growth type of the functions $\gamma_{x,\mathcal{U}}$ and $\gamma_{x,\mathcal{V}}$ is the same. If $\mathcal{F}$ is of class $C^1$ then the growth type of the functions $\gamma_{x,\mathcal{U}}$ is the same for all points $x$ is a same fixed leaf of $\mathcal{F}$.*

We define therefore the *combinatorial growth type* of a leaf $L$ of a $C^1$ foliation $\mathcal{F}$ on a compact manifold $M$ as the growth type of the function $\gamma_{x,\mathcal{U}}$ where $x \in L$ is any point and $\mathcal{U}$ is any finite cover of $M$ by distinguished neighborhoods. According to what we have seen, in this compact case, the growth type of a leaf $L$ of $\mathcal{F}$ is equal to the growth type of the orbit of any point $x \in L$ in the holonomy pseudogroup $\Gamma_{\mathcal{F}}$ of $\mathcal{F}$. In case the manifold is compact we also have:

**Proposition 9.4** ([Godbillon (1991)]). *Let $\mathcal{F}$ be a $C^1$ transversely oriented foliation on a oriented compact Riemannian manifold $M$. The (geometric) growth type of any leaf $L$ of $M$ is equal to the combinatorial growth type of $L$.*

### 9.1.5   Growth of groups

Let $G$ be a finitely generated abstract group. There exists therefore a subset $G^o \subset G$ such that any element $g \in G$ writes as $g = \prod_{\alpha \in A} g_\alpha^{n_\alpha}$ where $A$ is finite set, $n_\alpha \in \mathbb{Z}$ and $g_\alpha \in G^o, \forall \alpha \in A$. The *set of generators $G^o$ is symmetric* if $g \in G^o$ then $g^{-1} \in G^o$. We will assume that $G^o$ is symmetric. For any $n \in \mathbb{N}$ we define the subset $G_n$ of elements of $G$ which can be expressed as a word of length at most $n$ in the generators. That is,

$$G_n = \{g \in G, g = g_{\alpha_1} \circ ... \circ g_{\alpha_k}, k \leq n, \ g_{\alpha_j} \in G^o, \forall j = 1, ..., k\}.$$

The *growth function* of $G$ is $g(n) = \sharp G_n, n \in \mathbb{N}$. This definition can be extended as follows [Plante and Thurston (1976)]: Let $d$ be any left-invariant metric on $G$. We assume that $G$ is *discrete* so that for any $g \in G$ there exists $\epsilon_g > 0$ such that the metric ball $B_G(g; \epsilon_g) \subset G$ contains only the element $g$. The *growth function* of the pair $(G, d)$ is therefore defined as: for $t > 0$, $\gamma(t) = \sharp B_G(e; t)$ where $e \in G$ is the identity. If $\gamma(t) < \infty, \forall t \geq 0$ then we say that $\gamma$ is the growth function of $(G, d)$. In case $G$ has a symmetric set of generators $S$ we may consider any function $n_1 : S \to \mathbb{R}^+$ such that $n_1^{-1}(0, r] < \infty, \forall r \geq 0$. For any $g \in G$ we define the function $n(g) := \min\{\sum_{i=1}^{\ell} n_1(s_i), \ g = \prod_{i=1}^{\ell} s_i, s_i \in S\}$. Clearly $n(.)$ defines a left-invariant metric $d$ on $G$ by setting $d(g, h) := n(g^{-1}h)$. If $S = G$ then any left invariant metric on $G$ is obtained this way. Let us precise our main definition:

**Definition 9.4** ([Plante and Thurston (1976)]). The pair $(G, d)$ has *polynomial growth of degree $k$* if there exists a polynomial $p(x)$ of degree $k$ such that $\gamma(t) \leq p(t), \forall t \geq 0$, where $\gamma(.)$ is the growth function of $(G, d)$. We may also consider polynomials of the form $ax^\lambda, \ \lambda \geq 0, \lambda \in \mathbb{R}$.

**Proposition 9.5** ([Plante and Thurston (1976)]). *Let $S$ be a finite symmetric set of generators of $G$ and $n_1 : S \to \mathbb{R}^+, n_1 \equiv 1$. Denote by*

$n(.)$ *the metric above corresponding to $n_1 \equiv 1$. The growth of $G$ is polynomial with respect to some left-invariant metric $d$ if and only if, $(G, n)$ has polynomial growth.*

**Example 9.1.** Some examples of growth of groups are given below:
- A finitely generated abelian group has polynomial growth.
- A non-cyclic free group has exponential growth once we have $\sigma(n) = 2m \sum_{k=0}^{n-1} (2m-1)^k = \dfrac{m(2m-1)^n - 1}{m-1}$.
- Let $M^2$ be a closed surface orientable of genus $g \geq 2$ then $\pi_1(M)$ has exponential growth.

A situation of particular interest is the case of groups of polynomial growth, on which we have essential contributions of J. Plante, J. Wolf and J. Milnor (cf. [Wolf (1968)], [Milnor (1968)], [Plante (1975)]).
We have:

1. A nilpotent group of finite type has polynomial growth.

2. A solvable group of finite type $G$ which *does not* have a nilpotent subgroup of finite index has exponential growth. In case $G$ has polynomial growth then $G$ is *polycyclic* ($G$ is *polycyclic* if $G = G_k \rhd G_{k-1} \rhd \cdots \rhd G_0 = \{e\}$ with $G_k/G_{k-1}$ cyclic).

3. J. Tits has shown ([Tits (1972)]) that the converse of 1. is true: a group of finite type having polynomial growth has a nilpotent subgroup of finite index.

## 9.2 Holonomy invariant measures

Let $X$ be a Hausdorff topological space and $\Gamma$ a pseudogroup of local homeomorphisms of $X$. Denote by $\sigma_c(X)$ the ring of subsets of $X$ generated by the compact sets. A measure $\mu$ on $\sigma_c(X)$ is $\Gamma$-*invariant* if:
(i) $\mu$ is non-negative, finitely additive, finite on compact sets.
(ii) $\forall g \in \Gamma$ and any measurable set $A \subset \text{Dom}(g)$ we have $\mu(g(A)) = \mu(A)$.

Consider now the case $M$ is a $C^\infty$ manifold and $\mathcal{F}$ is a (regular) foliation of codimension $k \geq 1$ on $M$, assumed to be transversely oriented, of class $C^\infty$. The *holonomy pseudogroup* defined by $\mathcal{F}$ will be denoted by $\Gamma(\mathcal{F})$.

**Definition 9.5 (Plante, [Plante (1975)]).** A foliation $\mathcal{F}$ is said to have a *measure preserving holonomy*, or *holonomy invariant measure*, if its holonomy pseudogroup has a non-trivial invariant measure which is finite on compact sets. The *support* of an $\mathcal{F}$-invariant measure $\mu$ is the set of points $x \in M$ such that: given any $k$-dimensional disk transverse to $\mathcal{F}$, $D^k \pitchfork \mathcal{F}$, with $x \in \text{Int}(D^k)$, we have $\mu(D^k) > 0$. Since $\mu$ is $\mathcal{F}$-invariant, $\text{supp}(\mu)$ is closed and $\mathcal{F}$-invariant.

**Example 9.2.** Assume $\mathcal{F}$ has a closed leaf $L$ on $M$. Given any transverse section $\Sigma \subset M$ transverse to $\mathcal{F}$ we define a measure on $\Sigma$ as follows: $\forall A \subset \Sigma$, $\mu(A) := \sharp\{A \cap L\}$. This defines an holonomy invariant measure. Let now $\mathcal{F}$ be given by a closed holomorphic 1-form $\Omega$ on a complex manifold $M$. The holonomy pseudogroup is naturally a pseudogroup of translations $\Gamma(\mathcal{F}) \subset (\mathbb{C}, +)$ and any leaf of $\mathcal{F}$ has trivial holonomy. Any Borel measure on $\mathbb{C}$ which is invariant by translations is also $\Gamma(\mathcal{F})$-invariant. Another situation comes when $\mathcal{F}$ is real given by a closed $C^\infty$ $k$-form $\Omega$ on $M$. In this case, given any transverse section $\Sigma \subset M$, transverse to $\mathcal{F}$, the restriction $\Omega|_\Sigma$ is a volume element ($\mathcal{F}$ is transversely oriented) which is positive on open sets. The fact that $\Omega$ is closed implies that the induced transverse measure is $\mathcal{F}$-invariant. Assume now that $M, \mathcal{F}, \Omega$ are holomorphic. Using the complex structure $J \colon TM \to TM$ we may consider the real part and the imaginary part $\text{Re}(\Omega), \text{Im}(\Omega)$ of $\Omega$. Take $\omega = \text{Re}(\Omega) \wedge \text{Im}(\Omega)$, this is a $2k$-form real form, which is closed and defines $\mathcal{F}$ as a real codimension $2k$ foliation. Therefore the restriction $\omega|_{\Sigma^{2k}}$, where $\Sigma = \Sigma^{2k}$ is regarded as a $2k$-dimensional real submanifold transverse to $\mathcal{F}$, is positive on open sets and defines an $\mathcal{F}$-invariant transverse measure.

**Example 9.3.** It is a fairly well-known fact that a compact manifold $M$ supporting a *codimension one $C^1$ Anosov* flow $\varphi_t \colon M \to M$ has fundamental group with exponential growth [Thurston (1972)]. Such a result is not stated for codimension one holomorphic foliations (see [Ghys (1995)]). We shall consider an example of such a situation [Scardua (1997)]. Let $M$ be a compact complex manifold of dimension $n$ equipped with a closed holomorphic 1-form $\omega$ and $f \colon M \to M$ an automorphism such that $f^*\omega = \lambda.\omega$ for some $\lambda \in \mathbb{C} \setminus S^1$. We put $\Omega(x, t) := t.\omega(x)$ on $M \times \mathbb{C}^*$ so that $d\Omega = \eta \wedge \Omega$ for $\eta = \frac{dt}{t}$. The 1-form $\eta$ is closed and holomorphic in $M \times \mathbb{C}^*$ so that according to [Scardua (1997)] $\Omega = 0$ defines a codimension one holomorphic foliation $\tilde{\mathcal{F}}$ on $M \times \mathbb{C}^*$ which is transversely affine. $\tilde{\mathcal{F}}$ is non-singular provided that $\omega$ is non-singular on $M$. The action

$$\tilde{\varphi} \colon \mathbb{Z} \times (M \times \mathbb{C}^*) \to M \times \mathbb{C}^*, \; \tilde{\varphi}(n, (x, t)) = (f^n(x), \lambda^{-n}.t)$$

where $n \in \mathbb{Z}$, $(x, t) \in M \times \mathbb{C}^*$, is a locally free action generated by the automorphism $\varphi \colon M \times \mathbb{C}^* \to M \times \mathbb{C}^*$, $\varphi(x, t) = (f(x), \lambda^{-1}.t)$. Since

$\varphi^*\Omega(x,t) = \Omega(x,t)$ and $\varphi^*\eta = \eta$ it follows that $\tilde{\mathcal{F}}$ induces a codimension one transversely affine holomorphic foliation $\mathcal{F}$ of the quotient manifold $V^{n+1} = \tilde{M}/\tilde{\varphi} = \tilde{M}/\mathbb{Z}$. We apply this construction in a concrete situation: Take $A: \mathbb{C}^2 \to \mathbb{C}^2$ as the linear automorphism given by

$$A = \begin{pmatrix} 1 & 1 \\ 1 & 2 \end{pmatrix}.$$

Then $A$ has eigenvalues $\lambda_s = \frac{3-\sqrt{5}}{2}$ and $\lambda_u = \frac{3+\sqrt{5}}{2}$. The corresponding eigen-spaces are generated by $v_s = (2, 1 - \sqrt{5})$ and $v_u = (2, 1 + \sqrt{5})$ respectively. The stable linear foliation and the instable linear foliation are given by the 1-forms $\tilde{\omega}_s = 2dx + (1+\sqrt{5})dy$ and $\tilde{\omega}_u = (1+\sqrt{5})dx - 2dy$ respectively. Take $\tilde{\mathcal{F}}_u : \tilde{\omega}_u = 0$ on $\mathbb{C}^2$. We consider the action of the integer lattice $\mathbb{Z}^2$ on $\mathbb{C}^2$ obtained in the natural way and put $\tilde{M} = \mathbb{C}^2/\mathbb{Z}^2 = \mathbb{C}^* \times \mathbb{C}^*$. The map $A$ leaves $\mathbb{Z}^2$ invariant so that it induces an automorphism $F: \tilde{M} \to \tilde{M}$, which is indeed given by $F(z, w) = (zw, w^2)$ for coordinates $z = e^{2\pi i x}, w = e^{2\pi i y}$ on $\mathbb{C}^* \times \mathbb{C}^*$.

Now we consider the $\mathbb{Z}$-action on $\mathbb{C}^*$ given by $\psi: \mathbb{Z} \times \mathbb{C}^* \to \mathbb{C}^*, (n, t) \mapsto \lambda_s^{-n}.t$. Then $M = \mathbb{C}^*/\mathbb{Z} \times \mathbb{C}^*/\mathbb{Z} = \tilde{M}/\mathbb{Z}$ obtained this way is a compact surface equipped with an automorphism $f: M \to M$ induced by $F: \tilde{M} \to \tilde{M}$ indeed, $F(\lambda_s^{-1}.z, \lambda_s^{-1}.w) = \lambda_s^{-2}.(zw, w^2)$.

Now, the 1-form $\tilde{\omega} = \tilde{\omega}_u$ satisfies $A^*(\tilde{\omega}) = \lambda_s^{-1}.\tilde{\omega}$ and corresponds to a Darboux 1-form $\tilde{\omega} = (1+\sqrt{5})\frac{dz}{z} - 2\frac{dw}{w}$ on $\tilde{M} = \mathbb{C}^* \times \mathbb{C}^*$. Therefore, we have $F^*\tilde{\omega} = \lambda_s^{-1}.\omega$ and finally since $\psi^*\tilde{\omega} = \tilde{\omega}$ it follows that $\tilde{\omega}$ induces a closed holomorphic 1-form $\omega$ on $M$ with the property that $f^*(\omega) = \lambda_s^{-1}.\omega$. Thus, according to the above construction, the manifold $V^3 = M \times \mathbb{C}^*/\mathbb{Z}$ obtained by quotienting $M \times \mathbb{C}^*$ with the action of $\mathbb{Z}$ given by the action of $f$ on $M$ and of the homotheties $t \mapsto \lambda_s.t$ on $\mathbb{C}^*$, is a compact complex 3-manifold equipped with a transversely affine codimension one holomorphic foliation $\mathcal{F}$ coming from the linear unstable foliation $\mathcal{F}_u$ on $\mathbb{C}^2$. The foliation $\mathcal{F}$ exhibits exponential growth (for any metric on the compact manifold $V^3$) because $A^n$ expands $v_u.\mathbb{C}$ by a factor $\lambda_u^n$. On the other hand, [Scardua (1997)], the leaves of $\mathcal{F}$ on $V$ are dense, biholomorphic to $\mathbb{C}^* \times \mathbb{C}^*$ or to $(\mathbb{C}^*/\mathbb{Z}) \times \mathbb{C}^*$.

**Example 9.4.** Let $G$ be a Lie group which has polynomial growth in some left invariant metric. Let $\Phi: G \times M \to M$ be a locally free smooth action on a manifold $M$. There exists a Riemannian metric on $M$ which restricts to the $\Phi$-orbits as the induced metric coming from $G$. Thus $\Phi$ defines a foliation $\mathcal{F}$ on $M$, whose leaves have polynomial growth for this metric. For instance we may take any locally free holomorphic action $\Phi \colon \mathbb{C}^n \times M \to M$ where $M$ is a complex manifold and the euclidian metric on $\mathbb{C}^n$. The foliation by $\Phi$-orbits on $M$ has polynomial growth for a suitable metric on $M$. For $n = 1$ we have a *holomorphic flow* whose orbits have polynomial growth for a given metric on $M$.

**Example 9.5.** Here we complexify an original example in [Plante (1975)]. Let $G$ be a simply-connected complex Lie group and $H < G$ a closed (Lie) subgroup of (complex) codimension one. Given any discrete subgroup $\Gamma < G$ the group $H$ acts on the quotient $G/\Gamma$ by left translations generating a foliation $\mathcal{F}$ of codimension one. The leaves of $\mathcal{F}$ are the orbits of the above action. Since $G$ is simply-connected the universal covering $G \to G/\Gamma$ lifts $\mathcal{F}$ into a foliation $\tilde{\mathcal{F}}$ on $G$ whose leaf space is the Riemann surface $H \setminus G$. The exact homotopy sequence of the fibration $G \underset{H}{\to} H \setminus G$ shows that (for $H$ connected) the manifold $H \setminus G$ is simply-connected since $G$ is simply-connected. Therefore, $H \setminus G$ is either diffeomorphic to $\mathbb{C}P(1), \mathbb{C}$ or $\mathbb{D}$. Therefore the action of $\Gamma$ on $H \setminus G$ defines a global holonomy of $\mathcal{F}$ as a subgroup of $\mathrm{Diff}(N)$ for $N \in \{\mathbb{C}P(1), \mathbb{C}, \mathbb{D}\}$, so that this global holonomy group is either a subgroup of Moebius maps, affine maps or $\mathrm{SL}(2, \mathbb{R})$. If $\Gamma$ is uniform, that is, the quotient $G/\Gamma$ is compact, then $G$ is unimodular and the action of $G$ on $H \setminus G$ has an invariant measure iff $H$ is unimodular iff there exists a $\Gamma$-invariant measure. Therefore, when $G/\Gamma$ is compact $\mathcal{F}$ admits an invariant measure iff $H$ is unimodular.

The existence of holonomy invariant measures is a consequence of subexponential growth for the leaves as stated below:

**Theorem 9.1 (Plante, [Plante (1974)]).** *Let $\mathcal{F}$ be a $C^2$ foliation of codimension $k \geq 1$ on the compact manifold $M$. Assume that $\mathcal{F}$ exhibits a leaf $L$ having subexponential growth. Then there exists a nontrivial holonomy invariant measure $\mu$ for $\mathcal{F}$ which is finite on compact sets and which has support contained in the closure $\overline{L} \subset M$ of $L$.*

It is also known that if a *codimension one* (real) foliation of class $C^2$ on a compact manifold admits a non-trivial holonomy invariant measure then $\mathcal{F}$ has a leaf with polynomial growth [Plante (1974)]. *Is this also true for complex foliations?*

## 9.3 Plante's theorem

Let $\mathcal{F}$ be a $C^0$ codimension $k$ foliation of a closed Riemannian $n$-manifold $M$. Given an open subset $U \subset M$ we denote by $\mathcal{F}/U$ the restriction of $\mathcal{F}$ to $U$, namely, the foliation whose leaves are the connected components of the leaves of $\mathcal{F}$ intersected with $U$. Recall that the leaves of $\mathcal{F}/U$ are called the *plaques* of $\mathcal{F}$ in $U$. Denote by $D^r$ the closed unit disk in $\mathbb{R}^r$, $r \in \mathbb{N}$.

We shall call a cover $\mathcal{U} = \{U_i : i \in I\}$ of $M$ *nice* if the collection $Int(\mathcal{U}) = \{Int(U_i) : i \in I\}$ is an open cover of $M$ and the following properties hold for all $i, j \in I$:

(1) $U_i$ is diffeomorphic to $D^{n-k} \times D^k$ and $\mathcal{F}/U_i$ is given by the trivial codimension $k$ foliation $\{D^{n-k} \times * : * \in D^k\}$ of $D^{n-k} \times D^k$.

(2) Every plaque of $\mathcal{F}$ in $U_i$ intersects at most one plaque of $U_j$.

To any such cover $\mathcal{U}$ we can associate the disjoint union $X = \cup_{i \in I} X_i$ where each $X_i$ is the leaf space of $\mathcal{F}/U_i$. Clearly each $X_i$ is diffeomorphic to $D^k$ so $X$ is a compact metric space. By a plaque of $\mathcal{F}/\mathcal{U}$ we mean a plaque of $\mathcal{F}$ in $U_i$ for some $i \in I$. A chain of plaques is a finite collection of plaques $\{P_1, \cdots, P_s\}$ of $\mathcal{F}/\mathcal{U}$ satisfying $P_i \cap P_{i+1} \neq \emptyset$ for all $1 \leq i \leq s - 1$.

Given a leaf $L$ of $\mathcal{F}$, a nice cover $\mathcal{U}$, $x \in L \cap X$ and $n \in \mathbb{N}$ we define $L^n(x)$ as the set of all $y \in X$ for which there are $1 \leq s \leq n$ and a chain of plaques $\{P_1, \cdots, P_s\}$ such that $x \in P_1$ and $y \in P_s$. Clearly $\#L^n(x) < \infty$ where $\#$ denotes the cardinality operation.

**Definition 9.6.** We shall say that $L$ has *exponential growth* if there are a nice cover $\mathcal{U}$ and $x \in X \cap L$ such that

$$\liminf_{n \to \infty} \frac{\log(\#L^n(x))}{n} > 0.$$

Otherwise we say that $L$ has *subexponential growth*. If instead there are $p \in \mathbb{N}$, $K > 0$, a nice cover $\mathcal{U}$ and $x \in X \cap L$ such that

$$\#L^n(x) \leq Kn^p, \qquad \forall n \in \mathbb{N},$$

then we say that $L$ has *polynomial growth of degree* $\leq p$.

It turns out that the above definitions depend neither on the point $x \in L$ nor on the nice cover $\mathcal{U}$.

Now we state and prove the following result due to Plante [Plante (1975)]. Recall that the *first Betti number* of $M$ is the rank $\beta_1(M)$ of the free part of its first integer homology group.

**Theorem 9.2 (Plante [Plante (1975)]).** *Every transversely oriented $C^1$ codimension one foliation of a closed manifold $M$ exhibiting leaves of subexponential growth also exhibits leaves of polynomial growth of degree $\leq \max\{0, \beta_1(M) - 1\}$.*

**Proof.** Fix a foliation $\mathcal{F}$ and a leaf $L$ of $\mathcal{F}$ as in the statement. Then, there is a nice cover $\mathcal{U}$ and $x \in X \cap L$ such that

$$\liminf_{n \to \infty} \frac{\log(\#L^n)}{n} = 0$$

where we have written $L^n$ instead of $L^n(x)$ for simplicity.

Let us prove the identity

$$\liminf_{n \to \infty} \frac{\#L^{n+1} - \#L^n}{\#L^n} = 0.$$

Suppose that it fails. Then, there is $d > 0$ such that $\#L^{n+1} \geq (1+d)\#L^n$ for $n$ large. By induction we obtain $\#L^{n+k} \geq (1+d)^k\#L^n$ for $n$ large and $k \in \mathbb{N}$. Taking log and $\liminf_{k \to \infty}$ in this inequality we obtain $\log(1+d) \leq 0$ which is absurd since $d > 0$. This proves the identity. Then,

$$\liminf_{n \to \infty} \frac{\#L^{n+1} - \#L^{n-1}}{\#L^n} = \liminf_{n \to \infty} \frac{\#L^{n+1} - \#L^n}{\#L^n}$$

$$+ \liminf_{n \to \infty} \frac{\#L^n - \#L^{n-1}}{\#L^n} = 0$$

and so there is a sequence $n_i \to \infty$ such that

$$\lim_{i \to \infty} \frac{\#L^{n_i+1} - \#L^{n_i-1}}{\#L^{n_i}} = 0. \tag{9.1}$$

For all $i$ we put the normalized counting measure $\mu_i$ in $X$ supported on $L^{n_i}$, i.e.,

$$\mu_i(A) = \frac{\#(A \cap L^{n_i})}{\#L^{n_i}}.$$

Since $X$ is compact, and so the space of all Borel probability measures is compact in the weak* topology, we can assume that there is a Borel probability measure $\mu$ in $X$ to which the sequence $\mu_i$ converges. In the sequel we shall prove an invariant property for this measure.

Given $x \in X_i$ we denote by $P_i(x)$ the plaque of $\mathcal{F}$ in $U_i$ containing $x$. Denote by $Dom(\cdot)$ and $Rang(\cdot)$ the domain and range operations. Every pair $(i, j)$ with $U_i \cap U_j \neq \emptyset$ defines a $C^1$ map $\gamma_{ij} : Dom(\gamma_{ij}) \subset X_i \to X_j$,

$$Dom(\gamma_{ij}) = \{x \in X_i : P_i(x) \cap P_j(y) \neq \emptyset \text{ for some } y \in X_j\} \text{ and } \gamma_{ij}(x) = y.$$

Denote by $\Gamma$ the collection of all such maps. Notice that each $\gamma \in \Gamma$ is invertible with inverse $\gamma_{ji}$ whenever $\gamma = \gamma_{ij}$. Therefore $\gamma^{-1} \in \Gamma$ for all $\gamma \in \Gamma$.

We claim that $\mu$ is $\Gamma$-invariant, *i.e.*,

$$\int f d\mu = \int f d(\gamma_* \mu),$$

for all continuous map $f$ and all $\gamma \in \Gamma$, where $\gamma_* \mu$ is the pullback of $\mu$ under $\gamma$. Indeed, for all such $f$ and $\gamma$ we have,

$$\int f d\mu - \int f d(\gamma_* \mu) = \lim_{i \to \infty} \frac{1}{\#L^{n_i}} \left( \sum_{y \in L^{n_i}} f(y) - \sum_{y \in L^{n_i}} f(\gamma(y)) \right).$$

But

$$\sum_{y \in L^{n_i}} f(y) - \sum_{y \in L^{n_i}} f(\gamma(y)) = \sum_{y \in L^{n_i}} f(y) - \sum_{y \in \gamma^{-1} L^{n_i}} f(y)$$

$$= \sum_{y \in L^{n_i} \setminus \gamma^{-1} L^{n_i}} f(y) - \sum_{y \in \gamma^{-1} L^{n_i} \setminus L^{n_i}} f(y) = \sum_{y \in L^{n_i} \Delta \gamma^{-1} L^{n_i}} (\pm f(y)),$$

where $\Delta$ above denotes symmetric difference. Then,

$$\left| \int f d\mu - \int f d(\gamma_* \mu) \right| \leq \left( \sup_y |f(y)| \right) \cdot \lim_{i \to \infty} \frac{\#(L^{n_i} \Delta \gamma^{-1} L^{n_i})}{\#L^{n_i}}. \qquad (9.2)$$

Let us prove the inclusion

$$L^n \Delta \gamma^{-1} L^n \subseteq L^{n+1} \setminus L^{n-1}, \qquad \forall n \in \mathbb{N}, \quad \forall \gamma \in \Gamma.$$

Indeed, we have $L^n \subset L^{n+1}$ and $\gamma L^n \subset L^{n+1}$ by definition. If $y \in L^n \setminus \gamma^{-1} L^n$ then $y \in L^{n+1}$ and if $y \in L^{n-1}$ then we would have $\gamma(y) \in L^n$ and so $y \in \gamma^{-1} L^n$ which is absurd. Then $L^n \setminus \gamma L^n \subset L^{n+1} \setminus L^{n-1}$. Analogously we prove $\gamma L^n \setminus L^n \subset L^{n+1} \setminus L^{n-1}$ and the inclusion follows.

Since $L^{n-1} \subset L^{n+1}$, (9.2) and the previous inclusion imply

$$\left| \int f d\mu - \int f d(\gamma_* \mu) \right| \leq \left( \sup_y |f(y)| \right) \cdot \lim_{i \to \infty} \frac{\#L^{n_i+1} - \#L^{n_i-1}}{\#L^{n_i}} = 0$$

proving the claim.

Denote by $\mathcal{N}$ the *nerve* of $Int(\mathcal{U})$, namely, the collection of all subsets $J$ of $I$ such that either $J = \emptyset$ or $J \neq \emptyset$ and $\cap_{i \in J} U_i \neq \emptyset$ (we have written $U_i$ instead of $Int(U_i)$ for the skae of brevity). Clearly if $J \in \mathcal{N}$ every subset of $J$ belongs to $\mathcal{N}$ thus it the nerve an abstract finite simplicial complex. Its vertices are the elements of $\mathcal{U}$ and there is an edge (1-simplex) joining

$U_i$ to $U_j$ if and only if $U_i \cap U_j \neq \emptyset$. A chain of $\mathcal{N}$ is a finite ordered subset $C = (U_{i_1}, \cdots, U_{i_s})$ of $\mathcal{U}$ such that $U_{i_j} \cap U_{i_{j+1}}$ for all $1 \leq j < s$ (we then say that $s$ is the length of $C$). If additionally $U_{i_s} \cap U_{i_1} \neq \emptyset$ then we say that $C$ is a closed chain. Since $\mathcal{N}$ is finite there is $d > 0$ such that every pair of vertices can be connected by a chain of length at most $d$. To each closed chain $C$ it corresponds the homotopy class $[C]$ which form together the homotopy group $\pi_1(\mathcal{N})$. A standard trick in algebraic topology shows that the set $\mathcal{G}$ of homotopy classes corresponding to closed chains of length at most $3d$ generates $\pi_1(\mathcal{N})$. Moreover, the homotopy class of a closed chain of length at most $nd$ can be written as the product of at most $n$ elements of $\mathcal{G}$. Finally let us mention that for suitable $\mathcal{U}$ the homotopy groups $\pi_1(M)$ and $\pi_1(\mathcal{N})$ are homeomorphic (e.g. [Bott and Tu (1982)] p. 148 or [Rotman (1988)]). We shall assume hereafter that this is the case.

Next we use $\mu$ to construct a homomorphism $\Phi : \pi_1(\mathcal{N}) \to \mathbb{R}$. First observe that every closed path in $M$ is homotopic to a closed path of the form

$$c = \alpha_1 * \beta_1 * \alpha_2 * \beta_2 * \cdots * \alpha_{r-1} * \beta_{r-1} * \alpha_r * \beta_r$$

where $\alpha_i \cup \beta_i$ are contained in some $U_j \in \mathcal{U}$, $\alpha_i \subset X_j$ is transverse to $\mathcal{F}$ and $\beta_i$ is tangent to $\mathcal{F}$ for all $i \in \{1, \cdots, r\}$. Then, we can define

$$\Phi([C]) = \sum_{i=1}^{r} \pm \left( \mu(\alpha_i(0,1)) + \frac{1}{2}\mu(\alpha_i(\{0\})) + \frac{1}{2}\mu(\alpha_i(\{1\})) \right)$$

where $c$ is a closed path as above contained in $C$ and the signal $\pm$ above depends on whether the transversely orientation of $\mathcal{F}$ is preserved or reverted by $\alpha_i$. Denote by $p$ the rank of the finitely generated subgroup $\Phi(\pi_1(\mathcal{N}))$ of $\mathbb{R}$. Since $\pi_1(M)$ and $\pi_1(\mathcal{N})$ are isomorphic we have $p \leq \beta_1(M)$.

So, to prove the theorem, it remains to prove that every leaf intersecting the support of $\mu$ has polynomial growth of degree $\leq \max\{0, p-1\}$.

Previously let us remark that for every subgroup $G$ of the additive group $\mathbb{R}$, every base $\{g_1, \cdots, g_p\}$ of $G$ and every $\delta > 0$ there is a polynomial $Q(x)$ of degree $\leq \max\{0, p-1\}$ satisfying the property below:

$$\#\{g \in G \cap (-\delta, \delta) : g = n_1 g_1 + \cdots + n_p g_p, n_1, \cdots, n_p \in \mathbb{Z},$$
$$|n_1| + \cdots + |n_p| \leq n\} \leq Q(n), \tag{9.3}$$

for all $n \in \mathbb{N}$. To see it define the maps $f : G \to \mathbb{Z}^p$, $f(n_1 g_1 + \cdots + n_p g_p) = (n_1, \cdots, n_p)$, and $g : \mathbb{R}^p \to \mathbb{R}$, $g(x_1, \cdots, x_p) = x_1 g_1 + \cdots + x_p g_p$. Clearly $g$ is linear and a left inverse of $f$ therefore

$$\{g \in G \cap (-\delta, \delta) : g = n_1 g_1 + \cdots + n_p g_p, n_1, \cdots, n_p \in \mathbb{Z}, |n_1| + \cdots + |n_p| \leq n\} =$$

$$\{v \in \mathbb{Z}^p \cap g^{-1}((-\delta, \delta)) : \|v\| \leq n\}$$

where $\|(v_1, \cdots, v_p)\| = |v_1| + \cdots + |v_p|$. But now it is easy to find a polynomial $Q(x)$ of degree $\leq \max\{0, p-1\}$ satisfying

$$\#\{v \in \mathbb{Z}^p \cap g^{-1}((-\delta, \delta)) : \|v\| \leq n\} \leq Q(n).$$

This proves (9.3).

Next take a leaf $W$ intersecting the support of $\mu$ at some point $x$. Again we write $W^n$ instead of $W^n(x)$ for simplicity. Define $W_i^n = W^n \cap X_i$ and set $\nu_i(n) = \#W_i^n$ for all $i \in I$. Certainly we have $\#W^n = \sum_{i \in I} \nu_i(n)$. We concentrate in an specific $\nu_i(n)$ and set $W_i^n = \{x_1, \cdots, x_{\nu_i(n)}\}$. It follows from the definition that there are a positive integer $n_1 \leq n$ and a chain of plaques $(U_1, \cdots, U_{n_1})$ such that $x \in U_1$ and $x_1 \in U_{n_1}$. Take $x_j$ for some $2 \leq j \leq \nu_i(n)$ and a path $\alpha_j \subset X_i$ from $x_j$ to $x_1$. Again by definition there is another chain of plaques $(\overline{U}_1, \cdots, \overline{U}_{n_j})$ from $x$ to $x_j$ with $n_j \leq n$. Evidently $\overline{U}_1 = U_1$ and $\overline{U}_{n_j} = U_{n_1}$ so the chain $C_j = (\overline{U}_1, \cdots, \overline{U}_{n_j}, U_{n_1-1}, \cdots, U_1)$ is closed. Moreover, $\Phi([C_j]) = \mu(\alpha_j([0,1))$ thus $\Phi([C_j]) \neq \Phi([C_k])$ for all $1 \leq j < k \leq \nu_i(n)$ since $x$ belongs to the support of the $\Gamma$-invariant measure $\mu$.

Now choose $G = \Phi(\pi_1(\mathcal{N})$, a base $\{g_1, \cdots, g_p\}$ of $G$ and $\delta = \mu(X_i)$ (which is positive for $X_i$ intersects the support of $\mu$). It follows from the remark above that there is a polynomial $Q(x)$ of degree $\leq \max\{0, p-1\}$ satisfying (9.3). Notice that each $\Phi([C_j])$ belongs to $G \cap (-\delta, \delta)$ since $\alpha_j \subset X_i$. Moreover, $[C_j]$ is represented $C_j$ which is a closed chain of length $n_j + n_1 \leq 2n$ thus

$$[C_j] = \xi_1^j * \cdots * \xi_{l_j}^j$$

where $1 \leq l_j \leq \left[\frac{2n}{d}\right]$ (here $[\cdot]$ denotes integer part). Since $\{g_1, \cdots, g_p\}$ is a base of $G$, and each $\Phi(\xi_k^j)$ belongs to $G$, we can select integers $n_{k,s}^j$ with $1 \leq k \leq l_j$ and $1 \leq s \leq p$ such that

$$\Phi(\xi_k^j) = \sum_{s=1}^p n_{k,s}^j g_s$$

so

$$\Phi([C_j]) = \sum_{k=1}^{l_j} \Phi(\xi_k^j) = \sum_{s=1}^p \left( \sum_{k=1}^{l_j} n_{k,s}^j \right) g_s = \sum_{s=1}^p m_s^j g_s,$$

where

$$m_s^j = \sum_{k=1}^{l_j} n_{k,s}^j.$$

But $\mathcal{G}$ is a finite set, so there is an upper bound $O$ for the set of integers $\{|n_{k,s}^j|\}$. Then,

$$|m_s^j| \leq \max\{|n_{k,s}^j|\}l_j \leq Ol_j \leq O\left[\frac{2n}{d}\right]$$

so each $\Phi([C_j])$ belongs to

$$\left\{g \in G \cap (-\delta, \delta) : g = \sum_{s=1}^p n_s g_s, n_s \in \mathbb{Z}, \sum_{s=1}^p |n_s| \leq pO\left[\frac{2n}{d}\right]\right\}.$$

Applying (9.3) we conclude that

$$\nu_i(n) \leq Q\left(pO\frac{2n}{d}\right).$$

Then, we are done since $i \in I$ is arbitrary and $\#W^n = \sum_{i \in I} \nu_i(n)$.    $\square$

Since leaves of polynomial growth of degree $\leq 0$ are compact we obtain the following corollary.

**Corollary 9.1 (Plante, [Plante (1975)]).** *Every $C^0$ transversely oriented codimension one foliation exhibiting a leaf of subexponential growth on a closed manifold with first Betti number $\leq 1$ has a compact leaf.*

Chapter 10

# Currents, Distributions, Foliation Cycles and Transverse Measures

## 10.1 Introduction

The next chapters of this text are dedicated to some other topics in the Global Theory of Foliations. In this chapter, special attention is paid to the consequences of the Theory of Currents on foliated manifolds. We will therefore exploit aspects, already mentioned in the first part, of growth of leaves and of groups as well as the existence of invariant transverse measures and of foliation cycles for a given foliation. Despite its certain informality our approach and exposition aim to clear the key-points of some central results of the classical theory (e.g. the bijection between transverse invariant measures and foliation cycles and homological versions of Novikov's compact leaf theorem) allowing this way the link between the classical real framework and the so called "Complex World", where the foliations are frequently singular and therefore the ambient manifold may not be compact. After constructing the bases of the theory of currents and foliation cycles in the real case we address the problem of giving a non-geometrical (?) proof of Novikov's compact leaf theorem. The central idea/philosophy is that such a proof may be somehow adapted to the complex setting. References for these two parts should be essentially contained in the works of J. Plante, D. Sullivan, S. Schwartzmann, D. Ruelle, A. Haefliger (for the real classic part) and M. McQuillan, M. Brunella, for the existing complex part; and may be found in the end of this text ([McQuillan (1998)], [McQuillan (2001)], [Brunella (1997)], [Brunella (1999)], [Demailly (1997)]).

## 10.2    Currents

This section is inspired in the expositions of [Schwartz (1966)], [De Rham (1955)] and [Griffiths and Harris (1978)]. The study of currents associated to foliations has proved to be very useful in the comprehension of topological dynamical phenomena related to foliations (cf. [Sullivan (1976)], [Schwartz-mann (1957)], [Plante (1975)], [Haefliger (1981)] et al). In this chapter we try to illustrate some of these applications. We shall begin with the basic definitions which are involved, with motivations coming from particular situations already well-known. The first step is to introduce the concept of *current*. We denote by $C_c^\infty(\mathbb{R}^n)$ the vector space of the functions $C^\infty$ of compact support $f \colon \mathbb{R}^n \to \mathbb{R}$. Endow $C_c^\infty(\mathbb{R}^n)$, as usual, with the topology of the uniform convergence in compact sets (for $f$ and its derivatives of all orders). A *distribution* in $\mathbb{R}^n$ is then a linear functional $T \in (C_c^\infty(\mathbb{R}^n))^*$, that is, a linear application $T \colon C_c^\infty(\mathbb{R}^n) \to \mathbb{R}$ which is continuous in the $C^\infty$ topology in $C_c^\infty(\mathbb{R}^n)$. We denote now by $A_c^p(\mathbb{R}^n)$ the $\mathbb{R}$-vector space of differential $p$-forms of class $C^\infty$ and compact support in $\mathbb{R}^n$, equipped with topology inherited from $C_c(\mathbb{R}^n)$ in the natural way. Then $A_c^p(\mathbb{R}^n)$ is complete and we can consider its topological dual $\mathcal{D}^{n-p}(\mathbb{R}^n)$. In what follows we take $p + q = n$.

**Definition 10.1.** A *current* of *degree* $q$ on $\mathbb{R}^n$ is an element $C \in \mathcal{D}^q(\mathbb{R}^n)$. Thus, a current of degree $q$ on $\mathbb{R}^n$ is a linear continuous form on the space of differential forms of class $C^\infty$ and degree $p = n - q$ having compact support in $\mathbb{R}^n$. Also we shall say that $C$ is a current of *dimension* $p$.

### 10.2.1    *Examples*

Some basic examples are listed below:

**1.** A current of degree $n$ in $\mathbb{R}^n$ is simply a distribution in $\mathbb{R}^n$.

**2.** Let $N^p \subset \mathbb{R}^n$ be an oriented submanifold of $\mathbb{R}^n$. The integration along $N^p$ defines a current $C(\varphi) := \displaystyle\int_N \varphi, \quad \varphi \in A_c^p(\mathbb{R}^n)$ of dimension of dimension $p$.

**3.** Let $\psi = \sum_j \psi_j \, dx_j$ (in affine coordinates $(x_1, \ldots, x_n) \in \mathbb{R}^n$) be a differential $q$-form with locally integrable coefficients ($\psi_J \in L^1_{\text{loc}}(\mathbb{R}^n)$). To $\psi$ we

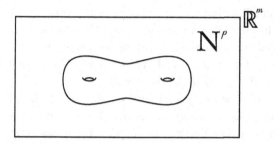

Fig. 10.1

can associate a current of degree $q$ (and dimension $p$) $C(\varphi) := \int_{\mathbb{R}^n} \varphi \wedge \psi$;
$\forall \varphi \in A_c^p(\mathbb{R}^n)$.

**4.** Given a singular $p$-chain $\alpha = \sum\limits_{j=1}^{r} a_j \cdot N_j$ in $\mathbb{R}^n$ we can (as in 2. above) define an integration current by setting

$$C(\varphi) := \int_\alpha \varphi, \quad \forall \varphi \in A_c^p(\mathbb{R}^n).$$

In $\{A_c^p(\mathbb{R}^n)\}$ we consider the exterior derivation of forms

$$d: A_c^p(\mathbb{R}^n) \to A_c^{p+1}(\mathbb{R}^n)$$
$$\varphi \mapsto d\varphi$$

and induce, in natural way, a derivation operator in $\mathcal{D}^q(\mathbb{R}^n)$:

$$d: \mathcal{D}^q(\mathbb{R}^n) \to \mathcal{D}^{q+1}(\mathbb{R}^n)$$
$$C \mapsto dC$$

$dC(\varphi) := C(d\varphi), \quad \forall \varphi \in A_c^{q+1}(\mathbb{R}^n)$.
In a natural way we obtain a complex of cochains $\{d: \mathcal{D}^p(\mathbb{R}^n) \to \mathcal{D}^{p+1}(\mathbb{R}^n)\}$ (naturally) associated to the complex of De Rham with compact support of $\mathbb{R}^n$

$$\{d: A_c^p(\mathbb{R}^n) \to A_c^{p+1}(\mathbb{R}^n)\}.$$

In particular, $d(dC) = 0$ for every current $C$ in $\mathbb{R}^n$.

We can "localize" the notions above in an obvious way: given open subset $U \subset \mathbb{R}^n$ we introduce the spaces $A_c^p(U)$ and $\mathcal{D}^q(U) := (A_c^p(U))^*$ where the topology we consider is the natural inherited from the topology of uniform convergence in compact parts (for functions and its derivatives of all orders) in $C_c^\infty(U)$. Given a diffeomorphism $C^\infty$ $F: U \to V$ between open subsets of $\mathbb{R}^n$ we have a natural homeomorphism $F^*\colon A_c^p(V) \to A_c^p(U)$ which is also linear. Thus, we can introduce the spaces of currents $\mathcal{D}^p(M^n)$ in a differentiable manifold $M^n$. Let us see properties of the corresponding complexes *of currents* $\{d\colon \mathcal{D}^q(M) \to \mathcal{D}^{q+1}(M)\}$ and of *De Rham* $\{d\colon A_c^p(M) \to A_c^{p+1}(M)\}$ in $M$. We recall:

A *complex of cochains* is a collection $\{d_k\colon A_k \to A_{k+1}\}_{k\in\mathbb{Z}}$ of abelian groups $A_k$ and group homomorphisms $d_k\colon A_k \to A_k$ with the property that $d_{k+1} \circ d_k = 0$. In particular we can consider the quotient groups

$$H^k := \frac{\mathrm{Ker}(d_k\colon A_k \to A_{k+1})}{\mathrm{Im}(d_{k-1}\colon A_{k-1} \to A_k)}$$

called the *Cohomology groups* of the complex considered. The *De Rham cohomology groups with compact support* of $M$ (differentiable manifold) denoted $H_{c,DR}^k(M)$ are defined this way (from $\{d\colon A_c^k(M) \to A_c^{k+1}(M)\}$) for $k \geq 1$ recalling that, by definition,

$$H_{c,DR}^0(M) := \{f\colon M \xrightarrow{C^\infty} \mathbb{R};\ f \text{ has compact support and } df = 0\}$$

is the number of compact connected components of $M$; also we have $H_{c,DR}^k(M) = 0$, $\forall k \geq n+1$ ($n = \dim M$) and we have the following:

$$H_{c,DR}^n(M;\mathbb{R}) = \begin{cases} \mathbb{R} & \text{if } M \text{ is orientable} \\ 0 & \text{if } M \text{ is non-orientable.} \end{cases}$$

**Remark 10.1.** We can also work with general differential forms (not necessarily with compact support) of class $C^\infty$ in $M$ obtaining the *De Rham complex* of $M$, whose cohomology is denoted by $H_{DR}^k(M)$, and the maximal order cohomology is given by:

$$H_{DR}^n(M) = \begin{cases} 0 & \text{if } M \text{ is non-compact or non-orientable} \\ \mathbb{R} & \text{if } M \text{ is compact and orientable.} \end{cases}$$

Let us return to the currents in $M$. As we have seen in the above examples, there exists a natural inclusion of the space of $q$-forms of class $C^\infty$ in $M$ in the space of currents of degree $q$ in $M$

$$\psi \mapsto C_\psi(\varphi) := \int_M \psi \wedge \varphi, \quad \forall \varphi \in A_c^p(M)$$

$\psi \in A^q(M)$ $q$-form $C^\infty$ in $M$.

Such inclusion gives indeed a homomorphism of complexes $\{i_p \colon A^q(M) \to \mathcal{D}^q(M)\}$ that induces by its turn a homomorphism in the cohomology groups

$$i_{p_\#} \colon H^q_{DR}(M) \to H^q(\mathcal{D}^*(M))$$

where $H^q(\mathcal{D}(M))$ denotes the order $q$ cohomology group of the complex of currents $\mathcal{D}^*(M)$ of $M$.

**Theorem 10.1 (Theorem of De Rham, [De Rham (1960)], [De Rham (1955)]]** Given a differentiable oriented manifold $M$ we have natural isomorphisms between the singular cohomology singular groups of Rham and of currents in $M$:

$$H^q_{\text{sing}}(M, \mathbb{R}) \simeq H^q_{DR}(M) \simeq H^q(\mathcal{D}^*(M)).$$

## 10.3 Invariant measures

Let $\mathcal{F}$ be a foliation of class $C^\infty$ dimension $p$ and codimension $q$ of a manifold $M^n$. There exists a cover $\mathcal{U} = \{U_j\}_{j\in\mathbb{N}}$ of $M$ with the following properties:

1. $\mathcal{U}$ is locally finite: given a compact $K \subset M$ we have $\#\{j \in \mathbb{N}; \ U_j \cap K \neq \emptyset\} < \infty$.

2. $U_j$ is connected and $\mathcal{F}|_{U_j}$ is trivial: there exists a diffeomorphism $\varphi_j \colon U_j \to \varphi_j(U_j) \subset \mathbb{R}^n$ such that $\varphi_j$ takes $\mathcal{F}$ onto the horizontal foliation in $\mathbb{R}^p \times \mathbb{R}^q = \mathbb{R}^n$.

3. In each $U_j$ we have an embedded disc $D^q \simeq \Sigma_j \subset U_j$ which is transverse to the plaques of $\mathcal{F}$ in $U_j$ and parametrizes this space of plaques.

We shall call $\mathcal{U}$ a *regular* cover of $M$ for the foliation $\mathcal{F}$. We also assume, with no loss of generality, that $\varphi_j(U_j) = \mathbb{R}^n$ and that $M = \bigcup_{j\in\mathbb{N}} \varphi_j^{-1}((-1,1)^n)$ and we can then take $\Sigma_j \subset \varphi_j^{-1}((-1,1)^n)$ and also rename $U_j = \varphi_j^{-1}[(-1,1)^n]$ in a way that:

4. Each leaf of $\mathcal{F}$ cuts some transverse disc $T_j$; and if $U_i \cap U_j \neq \emptyset$ then each plaque of $\mathcal{F}|_{U_i}$ meets at most one plaque of $\mathcal{F}|_{U_j}$ defining local diffeomorphisms $C^\infty$ say $g_{ij} \colon \Sigma_i \to \Sigma_j$ with the property that in $U_i \cap U_j$

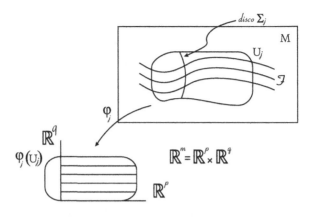

Fig. 10.2

we have $y_j = g_{ij} \circ y_i$ where $y_j = $ is the projection of $U_j$ onto $\Sigma_j$ (via the chart $\varphi_j$).

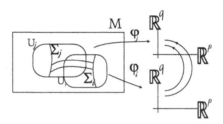

Fig. 10.3

Clearly we have the following condition of cocycle:

**5.** $U_i \cap U_j \neq \emptyset \Rightarrow g_{ij} = g_{ji}^{-1}$ and

$$U_i \cap U_j \cap U_k \neq \emptyset \Rightarrow g_{ij} \circ g_{jk} = g_{ik}$$

in the corresponding domains.

**Definition 10.2.** The *holonomy pseudogroup* of $\mathcal{F}$ for a regular cover $\mathcal{U}$ is the pseudogroup $\Gamma_{\mathcal{U}}$ of local diffeomorphisms $C^{\infty}$ of the manifold $\Sigma_{\mathcal{U}}$, disjoint sum of transverse discs $\Sigma_j$, generated by the local diffeomorphisms $g_{ij}$.

We recall the following definition:

**Definition 10.3.** Let $X$ be a topological space Hausdorff and $\Gamma$ a collection of local homeomorphisms $g \colon U \to V$ where $U, V \subset X$ are open subsets of $X$. Let us denote by $\mathrm{Dom}(g)$ and $\mathrm{Im}(g)$ the domain and the image of $g \in \Gamma$ respectively. We say then that $\Gamma$ is a *pseudogroup* of local homeomorphisms of $X$ if:

(i) $\forall g \in \Gamma$ we have $g^{-1} \in \Gamma$, $\mathrm{Dom}(g) = \mathrm{Im}(g^{-1})$ and $\mathrm{Im}(g) = \mathrm{Dom}(g^{-1})$;

(ii) If $g_1, g_2 \in \Gamma$ and $g \colon \mathrm{Dom}(g_1) \cup \mathrm{Dom}(g_2) \to \mathrm{Im}(g_1) \cup \mathrm{Im}(g_2)$ is a homeomorphism such that $g|_{\mathrm{Dom}(g_j)} = g_j$, $j = 1, 2$ then $g \in \Gamma$.

(iii) $\mathrm{Id} \colon X \to X$ belongs to $\Gamma$.

(iv) If $g_1, g_2 \in \Gamma$ then $g_1 \circ g_2 \in \Gamma$ com $\mathrm{Dom}(g_1 \circ g_2) \subset g_1^{-1}(\mathrm{Im}(g_2)) \cap \mathrm{Dom}(g_1)$.

(v) If $g \in \Gamma$ and $U \subset \mathrm{Dom}(g)$ is open then $g|_U \in \Gamma$.

Under these conditions we define the *orbit* of a point $x \in X$ in the pseudogroup $\Gamma$ by $\Gamma(x) := \{g(x) \in X, g \in \Gamma \text{ and } x \in \mathrm{Dom}(g)\}$.

We denote by $\sigma_c(X)$ the ring of subsets of $X$ generated by the compact sets. A measure $\mu$ in $\sigma_c(X)$ is said to be $\Gamma$-*invariant* if:

(vi) $\mu$ is non-negative, finitely additive, and finite in compact sets.

(vii) $\forall g \in \Gamma$ and any measurable subset $A \subset \mathrm{Dom}(g)$ we have $\mu(g(A)) = \mu(A)$.

In the above case, of the holonomy pseudogroup of the foliation $\mathcal{F}$ relative to the regular cover $\mathcal{U}$ we conclude that, in fact, $\Gamma_{\mathcal{U}}$ is a pseudogroup of local diffeomorphisms $C^\infty$ of $\Sigma_{\mathcal{U}}$. In case we have another regular cover of $M$ relative to $\mathcal{F}$, say $\widetilde{\mathcal{U}} = \{\widetilde{U}_j\}_{j \in \mathbb{N}}$ if we suppose that $\widetilde{\mathcal{U}}$ is *thinner* than $\mathcal{U}$, (*i.e.*, for each index $j \in \mathbb{N}$ there exists an index $\nu(j) \in \mathbb{N}$ such that $\widetilde{U}_j \subseteq U_{k(j)}$), and also that $\widetilde{U}_j$ is *uniform* in $U_{k(j)}$ (*i.e.*, each plaque of $U_{k(j)}$ meets *at most* one plaque of $\widetilde{U}_j$) then we obtain a natural identification between the corresponding holonomy pseudogroups $\Gamma_{\widetilde{\mathcal{U}}} \xrightarrow{\sim} \Gamma_{\mathcal{U}}$. This shows the following (exercise!):

**Lemma 10.1.** *All the holonomy pseudogroups $\Gamma_{\mathcal{U}}$, where $\mathcal{U}$ is cover regular of $M$ for a foliation $\mathcal{F}$, are naturally equivalent.*

We may then introduce the (well-defined whether $M$ is compact or not) *holonomy pseudogroup of the foliation $\mathcal{F}$*. This way we can formalize the following notion:

**Definition 10.4.** A foliation $\mathcal{F}$ of a manifold $M$ is said to admit a *holonomy invariant transverse measure* (or simply *invariant transverse measure*) if its holonomy pseudogroup has some invariant measure (non-trivial) which is finite in compact sets. The *support* of an invariant measure $\mu$ is the set of points $x \in M$ such that: given any transverse disc to $\mathcal{F}$ of dimension $q = $ codimension of $\mathcal{F}$, $D^q \subset M$ with $x \in \text{Int}(D^q)$, we have $\mu(D^q) > 0$. The support of $\mu$, denoted by $\text{supp}(\mu)$, is closed and (since $\mu$ is invariant) it is saturated (invariant) by $\mathcal{F}$.

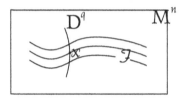

Fig. 10.4

Let us see some examples illustrating the notions above:

### 10.3.1   *Examples*

Let us now see some examples of foliations and invariant transverse measures.

**1.** Let $\mathcal{F}$ be a foliation of codimension 1 given by a non-singular closed 1-form of class $C^\infty$, $\Omega$ in $M$. Then it is to see from the Poincaré Lemma that the holonomy pseudogroup of $\mathcal{F}$ is naturally a group of translations $\Gamma_{\mathcal{F}} \subset (\mathbb{R}, +)$ and any leaf of $\mathcal{F}$ has trivial holonomy group (a translation with a finite fixed point is the identity). Therefore, any Borel measure in $\mathbb{R}$ invariant by translations is also $\Gamma_{\mathcal{F}}$-invariant. Suppose now that $\mathcal{F}$ is of codimension $k$ and given by a closed $k$-form $\Omega$ in $M$. In this case given any transverse $k$-disc to $\mathcal{F}$ say $D^k \subset M$, the restriction $\Omega|_{D^k}$ is a volume form (assume also that $\mathcal{F}$ is transversely oriented this way) which is positive in open sets. The fact that $\Omega$ is closed implies that the transverse measure this way induced is $\mathcal{F}$-invariant.

**2.** Suppose that $\mathcal{F}$ admits a closed leaf $L_0 \subset M$. Given a transverse disc to $\mathcal{F}$, say $D \subset M$ we define for any $A \subset D$ a measure $\mu(A) := \sharp\{A \cap L\}$. Clearly we obtain this way an invariant transverse measure $\mu$ for $\mathcal{F}$; also we have $\operatorname{supp}(\mu) = L_0$.

**3.** Let $\mathcal{F}$ be a foliation defined by the fibration $M \to B$ of $M$ over a manifold $B$; then the transverse measures for $\mathcal{F}$ correspond to the measures over $B$, which are finite on compact sets.

**4.** Let $\mathcal{F}$ be a foliation of $M$ and $f \colon \widetilde{M} \to M$ a proper application of class $C^\infty$ and transverse to $\mathcal{F}$. Denote by $\widetilde{\mathcal{F}}$ the lift $f^*\mathcal{F}$ to $\widetilde{M}$, if $\mathcal{F}$ admits an invariant transverse measure $\mu$ then $\widetilde{\mathcal{F}}$ admits an invariant transverse measure $\widetilde{\mu} := f^*(\mu)$ defined naturally by $\widetilde{\mu}(\widetilde{D}) := \mu(D)$, where $\widetilde{D} = f^{-1}(D)$ as in the figure below.

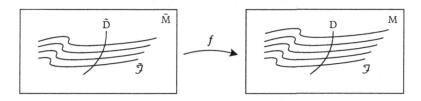

Fig. 10.5

**5.** According to Joe Plante [Plante (1975)] "if $\mathcal{F}$ is foliation of class $C^2$ of a *compact* manifold $M$ admitting a leaf $L_0$ with subexponential growth (geometrical, Riemannian) then $\mathcal{F}$ admits an invariant transverse measure $\mu$, finite in compact sets, whose support $\operatorname{supp}(\mu) \subset \overline{L}_0$."

6. Let us consider now more in details the case of suspensions: Let $\pi \colon E \xrightarrow{F} B$ be a fiber bundle of class $C^\infty$ with typical fiber $F$; base $B$, projection $\pi$ and total space $E$. We say that a foliation $\mathcal{F}$ of $E$ is *transverse to the fibers* of $E$ if:

(a) Given $x \in E$ we have which $L_x$ is transverse to the fiber $F_{\pi(x)} = \pi^{-1}(\pi(x))$ and in fact $\dim \mathcal{F} + \dim F = \dim E$.

(b) The restriction $\pi|_L \colon L \to B$, where $L$ is an arbitrary leaf of $\mathcal{F}$, is a covering map.

**Remark 10.2.** We observe that if the fiber $F$ is compact then (b) follows from (a); even for $B$ non-compact (see [Camacho and Lins-Neto (1985)] page 94).

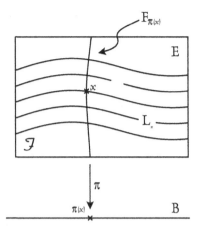

Fig. 10.6

Since each restriction $\pi|_L \colon L \to B$ is a covering map we can define a representation $\varphi \colon \pi_1(B) \to \mathrm{Dif}^\infty(F)$ from the fundamental group of the base $B$ into the group of (global) diffeomorphisms of class $C^\infty$ (suppose $\mathcal{F}$ of class $C^\infty$), of the fiber $F$ as follows: Fixed base points $b_0,\ b_0' \in B$. Given a path $\alpha \colon [0,1] \to B$, $\alpha(0) = b_0$, $\alpha(1) = b_0'$ we define, for each $y \in F_{b_0}$, the point $f_\alpha(y) \in F_{b_0}$, as the final point final $\tilde{\alpha}_y(1)$ of the lift $\tilde{\alpha}_y \colon [0,1] \to L_0$ of $\alpha$, by the covering map $\pi|_{L_y} \colon L_y \to B$, with origin at the point $y = \tilde{\alpha}_y(0)$.

For $b_0 = b_0'$ we identify $F_{b_0} = F_{b_0'} \simeq F$ and we obtain representation

$$\varphi \colon \pi_1(B, b_0) \to \mathrm{Dif}^\infty(F)$$
$$[\alpha] \mapsto f_{[\alpha]}\,.$$

The image of this representation is called *global holonomy of* $\mathcal{F}$. By means of a constructive process one may prove the following

"I. Let $\mathcal{F}$ and $\mathcal{F}'$ be foliations transverse to the fibers of a fiber bundle $\pi \colon E \xrightarrow{F} B$. Then the groups of global holonomy of $\mathcal{F}$ and $\mathcal{F}'$ are conjugate (in $\mathrm{Dif}^\infty(F)$) if and only if $\mathcal{F}$ and $\mathcal{F}'$ are conjugate by a fibred diffeomorphism $\psi \colon E \to E$.

II. Given a fiber bundle space $\pi \colon E \xrightarrow{F} B$ there exists a foliation $\mathcal{F}$ transverse to the fibers of the bundle if and only if the structural group of

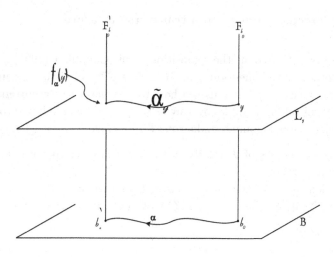

Fig. 10.7

the bundle is discrete.

III. Given a representation $\varphi\colon \pi_1(B) \to \mathrm{Dif}^\infty(F)$ of the fundamental group of a manifold $B$ in the group of $C^\infty$ diffeomorphisms of a manifold $F$ with image $G < \mathrm{Dif}^\infty(F)$ there exist a foliation $\mathcal{F}$ of class $C^\infty$ of the manifold $E$, a structure of fiber bundle space $\pi\colon E \xrightarrow{F} B$ such that $\mathcal{F}$ is transverse to the fibers of the bundle and whose global holonomy (of $\mathcal{F}$) is conjugate to $G$. By (I) $\mathcal{F}$ is unique up to natural equivalence."

We shall call such a foliation $\mathcal{F}$ the *suspension* of the representation $\varphi\colon \pi_1(B) \to \mathrm{Dif}^\infty(F)$.

We recall that a group $G$ is *amenable* if the space $\mathcal{B}(G) := \{f\colon G \to \mathbb{R}; \|f\|$ is bounded$\}$, equipped with the norm of the supreme, admits a positive linear functional $\xi\colon \mathcal{B}(G) \to [0, +\infty)$ with $\xi(1) = 1$ and $G$-invariant, (*i.e.*, $\xi(f \circ L_g) = \xi(f)$, $\forall f \in \mathcal{B}(G)$. Such a functional $\xi$ is called *continuous invariant mean* (cf. [Hirsch (1971)]). It is proven that if $G$ is a finitely generated group and with subexponential growth then $G$ is *amenable* (cf. [Thurston (1974)]) and that solvable finitely generated groups (for example) are amenable.

Let finally $\mathcal{F}$ be a foliation transverse to the fibers of a fiber bundle $\pi\colon E \xrightarrow{F} B$ and suppose that the group of global holonomy of $\mathcal{F}$ is amenable then, if a fiber $F$ is compact, $\mathcal{F}$ has invariant transverse measures.

## 10.4   Current associate to a transverse measure

This section is based in the exposition from [Godbillon (1991)]. Let $\mathcal{F}$ be a foliation of codimension $q$ in $M^n$, class $C^\infty$, admitting an invariant transverse measure $\mu$. Let us see how to associate to $\mu$ a current $C_\mu$ in $M$; we begin taking regular covering $\mathcal{U} = \{U_j\}_{j \in \mathbb{N}}$ of $M$ relative to $\mathcal{F}$ and considering the holonomy pseudogroup $\Gamma_\mathcal{U}$. As before $\Sigma_\mathcal{U} = \bigcup_{j \in \mathbb{N}} \Sigma_j$ denotes the space of plaques of $\mathcal{F}$ relatively to $\mathcal{U}$; we can then disintegrate the a measure $\mu$ as follows:

- $\mu$ defines a Borelian measure over $\Sigma_\mathcal{U}$ invariant by $\Gamma_\mathcal{U}$.
- Let $\sum_{j \in \mathbb{N}} a_j = 1$ be a partition $C^\infty$ of the unity, strictly subordinate to the cover $\mathcal{U}$ of $M$.
- Given $\varphi \in A_c^{n-q}(M)$ of class $C^\infty$ and degree $n - q$ ($n = \dim M$) and compact support in $M$ we can consider the product $a_j\varphi \in A_c^{n-q}(\Sigma_j)$ as a continuous function on $\Sigma_j$ provided that $\mathcal{F}$ is oriented and we consider in $U_j$ (and therefore in $\Sigma_j$) the orientation induced by $\mathcal{F}$. In fact, we can consider a function $y \mapsto \int_{P_y} a_j\varphi$

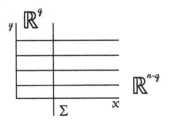

Fig. 10.8

defined in terms of local coordinates $(x, y)$ in $U_j$ that make $\mathcal{F}|_{U_j}\{y = \text{cte}\}$; the plaques of $\mathcal{F}|_{U_j}$ are of the form $P_y \subset \mathbb{R}^{n-q} \times \{y\}$ and a transverse $\Sigma_j$ of the form $\Sigma_j \subset \{x = 0\}$.

- We integrate and sum these functions obtaining the value

$$C_\mu(\varphi) := \sum_{j \in \mathbb{N}} \int_{\Sigma_j} \left( \int_{P_y} a_j\varphi(y) \right) d\mu(y).$$

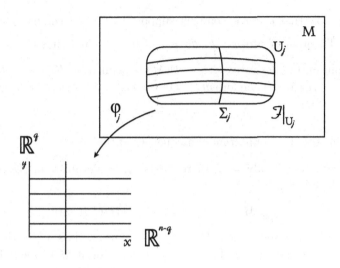

Fig. 10.9

Using the fact that $\mu$ is invariant by the local diffeomorphisms $g_{ij} : \Sigma_i \to \Sigma_j$ we conclude that in fact the value of $C_\mu(\varphi)$ does not depend on the $\sum\limits_{j \in \mathbb{N}} a_j = 1$ (partition of the unity) neither on the regular cover $\mathcal{U}$ with respect to $\mathcal{F}$ (there is no need to suppose $M$ compact).

**Definition 10.5.**  $C_\mu$ is the *current associate* to the invariant transverse measure $\mu$ for $\mathcal{F}$.

The following result is central in the theory:

**Proposition 10.1.** $C_\mu$ *is a closed current.*

**Proof.** Using the above notations we have that

$$C_\mu(d\varphi) = \sum_{j \in \mathbb{N}} \int_{\Sigma_j} \left( \int_{P_y} (a_j \, d\varphi)(y) \right) d\mu(y).$$

On the other side, $d\varphi = d\left( \sum_j a_j \varphi \right) = \sum_j d(a_j \varphi)$ so that

$$C_\mu(d\varphi) = \sum_{j \in \mathbb{N}} \int_{\Sigma_j} \left( \int_{P_y} d(a_j \, d\varphi)(y) \right) d\mu(y).$$

Assume now that $a_j\varphi$ has compact support in $P_y$ (not compact) so that, by the Theorem of Stokes, $\displaystyle\int_{P_y} d(a_j\varphi) = 0$, $\forall y$ and thus $C_\mu(d\varphi) = 0$. By definition the derivative $dC_\mu \in \mathcal{D}(M)$ is defined by $dC_\mu(\varphi) := C_\mu(d\varphi)$ where $\varphi \in A_c(M)$. Therefore $dC_\mu = 0$, that is, $C_\mu$ is a closed current.

$\square$

Let us see some important consequences of this result:

As we have already seen, there exists an isomorphism of (De Rham) cohomology groups

$$H^q_{\text{sing}}(M, \mathbb{R}) \simeq H^q_{DR}(M) \simeq H^q(\mathcal{D}^*(M));$$

therefore each closed current $C \in \mathcal{D}^q(M)$ defines a class $[C]$ in the space $H^q_{DR}(M)$. By the Duality Theorem of of Poincaré, if $M$ is orientable, we have a natural isomorphism $H^q_{DR}(M) \simeq \left(H^p_{c,DR}(M)\right)' =$ topological dual space of the cohomology group (of degree $p$) with compact support in $M$, of De Rham. Hence, we can associate to $C$ a class $[[C]]$ in

$$\left(H^p_{c,DR}(M)\right)' \simeq \begin{cases} H_p(M, \mathbb{R}) & \text{if} \quad M \quad \text{is compact} \\ H^q(M, \mathbb{R}) & \text{if} \quad M \quad \text{is orientable.} \end{cases}$$

Thus, for $M$ compact (respectively orientable) we have associate an invariant transverse measure for $\mathcal{F}$, the *homology class* (respectively *class of cohomology*) *of this measure.*

Let us see some examples:

### 10.4.1  *Examples*

**1.** If $N^p \subset M^n$ is an oriented submanifold compact of dimension $p$ invariant by $\mathcal{F}$ then the class of the current of integration corresponding to $N$ is the class $[N]$ of homology of $N$ in $H_p(M, \mathbb{R})$; note that $N$ is a compact leaf of $\mathcal{F}$.

**2.** Let $\mathcal{F}$ be a foliation of dimension $p$ and codimension $q$ of $M^n$. Assume that $\mathcal{F}$ and $M$ are oriented and that is $\mathcal{F}$ transversally oriented. The differential form $\Omega$ of degree $q$ in $M$ such that for each transverse disc to $\mathcal{F}$, $D^q \subset M$ we have $\Omega|_{D^q}$ is the form of volume (positive for the induced orientation in $D^q$) is a *transverse volume form* of $\mathcal{F}$ in $M$. We can choose a continuous vector field $X_{\mathcal{F}}$ of $p$-vectors on $M$ such that in each point $x \in M$ we have $T_x\mathcal{F} = $ oriented space generated by $X_{\mathcal{F}}(x)$.

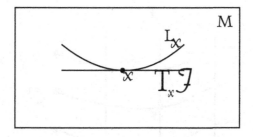

Fig. 10.10

In this case we can obtain a transverse volume form positive $\nu_{\mathcal{F}}$ for $\mathcal{F}$ in $M$ of class $C^\infty$ such that $\nu_{\mathcal{F}}(X_{\mathcal{F}}) = 1$ in $M$. We shall say that $\nu_{\mathcal{F}}$ is *normalized* for $X_{\mathcal{F}}$. In a general way, given transverse volume $q$-form $\Omega$ for $\mathcal{F}$ in $M$ the associated current to $\Omega$ is defined by $C(\varphi) = \displaystyle\int_M \Omega \wedge \varphi$ and the homology class corresponding to $C$ in $H^q(M, \mathbb{R})$ is the corresponding class of $\Omega$ in $H^q_{DR}(M)$.

**3.** Let $\alpha = \sum_{j=1}^{r} a_j N_j$, $a_j \in \mathbb{Z}$; be a singular $p$-chain in $M^n$ and denote by $C$ the current of integration defined by $\alpha$ in $M$; if $\alpha$ is closed ($\partial\alpha = 0$) then $C$ is closed ($dC = 0$) as consequence of the Theorem of Stokes. The class of $C$ in $H_p(M, \mathbb{R})$ is the class of $\alpha$ in this same space. In this example we are not necessarily assuming the existence of a foliation in $M$ which leaves $\alpha$ invariant.

**4.** Let now $\mathcal{F}$ be a foliation transverse to the fibers of the bundle $\pi\colon E \xrightarrow{F} B$ in $E$; given an invariant transverse measure $\mu$ we have that $\mu$ corresponds (in bijective way) to a Borelian measure $\mu_0$ over the fiber $F$ which is invariant by the global holonomy $\mathrm{Hol}(\mathcal{F}) \subset \mathrm{Dif}(F)$ of $\mathcal{F}$, and finite in compact sets of $F$. Let $C$ be a current corresponding a $\mu$; then by construction we have which

$$C(\varphi) = \int_B \left( \int_{F_{\pi(y)}} \varphi\, d\mu(y) \right) = \int_B f_\varphi(y) d\mu_0(y)$$

where $f_\varphi\colon B \to \mathbb{R}$ is defined by the integration of $\varphi$ along the fibers (cf. the figure below).

Thus, in order to study the class of $C$ in $H^q(M, \mathbb{R})$ we can report to the (class of the) measure $\mu_0$ in $H^*(B, \mathbb{R})$.

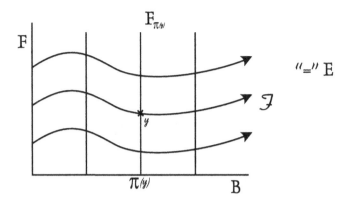

Fig. 10.11

Suppose now that the fiber $F$ is compact and let us study the homology class of (one fiber) $[F_0]$ in $H_q(M, \mathbb{R})$. Take a tubular neighborhood $\Gamma \colon W \to F_0$ of this fiber in $E$ such that the projection $\Gamma$ has as fibers (transverse discs) the leaves of $\mathcal{F}|_W$.

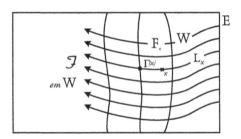

Fig. 10.12

We can assume $\overline{W} \subset M$ compact and using "bump functions" we obtain closed form $\varphi \in A_c^p(M)$ such that supp $\varphi \subset W$ and $\displaystyle\int_{D_x} \varphi = 1, \quad \forall x \in F_0$ where $D_x = r^{-1}(x)$ is a fiber of $\Gamma$ by $x \in F_0$.

But then we have $C(\varphi) = \mu_0(\pi(W)) > 0$ so that $C(\varphi) \neq 0$. On the other hand, since $\varphi$ is closed we have that its class in $H_{c,DR}^p(M, \mathbb{R})$ is dual to the class of $F_0$ in $H_q(M, \mathbb{R})$ so that if $[F_0] = [0]$ in $H_q(M, \mathbb{R})$ then $[\varphi] = [0]$ in

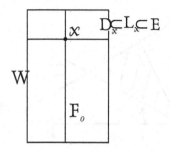

Fig. 10.13

$H^p_{c,DR}(M, \mathbb{R})$ and so $C(\varphi) = 0$ giving a contradiction. This shows that "the class of $F_0$ is not zero in $H_q(M, \mathbb{R})$". Since $F_0$ is arbitrary we conclude the same for any fiber of $\pi \colon E \to B$. The same proof gives us:

**6.** "Let $\mathcal{F}$ and $M$ be oriented and $N^q \subset M^n$ compact submanifold without boundary and transverse to $\mathcal{F}$. If there exists invariant transverse measure $\mu$ for $\mathcal{F}$ with supp $\mu \cap N \neq \emptyset$ then $[N] \neq [0]$ in $H_q(M, \mathbb{R})$.

## 10.5  Cone structures in manifolds

In this section we shall follow [Sullivan (1976)]. Let $\mathbb{E}$ be a real locally convex topological vector space. Given a convex cone $C \subset \mathbb{E}$ we say that $C$ is a connected convex *compact* if there exists linear functional $\varphi \colon \mathbb{E} \to \mathbb{R}$ such that

1.  $\varphi(x) > 0, \quad \forall x \in C \backslash \{0\}$.
2.  $\varphi^{-1}(1) \cap C$ is compact; called the *base* of the cone.

We denote by $\overset{\circ}{C}$ the set of radii of of $C$; $\overset{\circ}{C}$ is direct identification with its base.

**Definition 10.6.** A *cone structure* in a closed subset $F$ of a manifold $C^\infty$  $M$ is a continuous field of convex compact sets cones, say $\{C_x\}_{x \in F}$, in the vector spaces $\mathfrak{X}_p(x)$ of tangent $p$-vectors in $M$ (for $x \in F$).

The continuity of the field $\{C_x\}_{x \in F}$ is defined in terms of the movement of its bases $\overset{\circ}{C}_x$ for a suitable metric in the radii (see [Sullivan (1976)]). Thus the $p$-form $C^\infty$  $\Omega$ in $M$ is said to be *transverse* to the cone structure

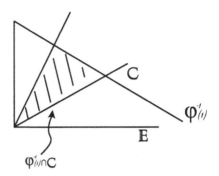

Fig. 10.14

$\{C_x\}_{x \in F}$ if $\Omega(x)(v_1, \ldots, v_p) > 0$, $\forall (v_1, \ldots, v_p) \in C_x \subset \mathfrak{X}_p(x)$ non-zero and $\forall\, x \in F$. Such transverse forms can always be constructed and determine currents: given point $x \in M$ we define the *Dirac current* associate to the fixed $p$-vector $X(p) \in \mathfrak{X}_p(x)$ by $\delta_{X,x} \colon \varphi \mapsto \varphi(X)(x)$; by choosing $X(x)$ in the cone $C_x \subset \mathfrak{X}_p(x)$ we obtain a collection of such Dirac currents which gives us a closed convex cone of currents called *cone of currents of structure* associate to the cone structure $\{C_x\}_{x \in F}$. When $F$ is compact the cone of currents of structure associate is a compact convex cone. We shall call the *structural cycles* of a cone structure in a manifold the structural currents which are closed (in the sense of currents). It is proven (cf. [Sullivan (1976)] §2) that if $F \subset M$ is compact then any structural current $C$ writes as $C = \displaystyle\int_F f\, d\mu$ where $\mu$ is a measure $\geq 0$ in $F$ and $f$ is an integrable function $\mu$-integrable taking values in $\mathfrak{X}_p(M) = \{p\text{-vectors in } m\}$ and such that $f(x) \in C_x$ (cone structure given originally, $\forall\, x \in F$).

# Chapter 11

# Foliation cycles: A homological proof of Novikov's compact leaf theorem

Let $\mathcal{F}$ be an oriented foliation of class $C^\infty$, dimension $p$ and codimension $q$ in $M$ oriented, $X_{\mathcal{F}}$ a continuous field of $p$-vectors generating $T\mathcal{F}$ and $\nu_{\mathcal{F}}$ transverse volume form normalized for $X_{\mathcal{F}}$. Clearly $\mathcal{F}$ defines (via $X_{\mathcal{F}}$) a *foliation current* of dimension $p$ over $M$; for each $x \in M$ we denote by $C_{\mathcal{F}}(x)$ the convex cone in $T_x M$ generated by the fields of $p$-vectors tangent to $\mathcal{F}$ in $x$ and denote by $C_{\mathcal{F}}$ the cone structure over $M$ obtained this way; an element of the cone of currents of structure associate to $C_{\mathcal{F}}$ is called a *foliation current* of $\mathcal{F}$. In other words, a foliation current of $\mathcal{F}$ is an element do convex cone closed do space of currents of dimension $p$ over $M$ which is generated by Dirac currents of the form $\delta_{X,x} \colon \varphi \mapsto \varphi(X)(x)$ where $x \in M$ and $X$ is a $p$-field tangent to $\mathcal{F}$.

**Definition 11.1.** A *foliation cycle* of $\mathcal{F}$ in $M$ is a foliation current of $\mathcal{F}$ which is closed in the sense of currents, that is, a structural cycle of $C_{\mathcal{F}}$.

Owing to the preceding chapter if $\mu$ is an invariant transverse measure for $\mathcal{F}$ in $M$ then a current associate to $\mu$ is a foliation cycle of $\mathcal{F}$. The converse in the compact case was proven by D. Sullivan (cf. [Sullivan (1976)]):
"Let $\mathcal{F}$ be a foliation $C^\infty$ of $M$ compact and suppose $\mathcal{F}$ and $M$ oriented. Then each foliation cycle for $\mathcal{F}$ in $M$ comes (via the construction already presented) of a unique invariant transverse measure for $\mathcal{F}$."
We define the support of a current in the obvious way and we can then observe that if $C$ is a foliation cycle for $\mathcal{F}$, coming from an invariant transverse measure $\mu$ in $M$, then $\operatorname{supp}(C) = \operatorname{supp}(\mu) \subset M$; in particular $\operatorname{supp}(C)$ is closed and $\mathcal{F}$-invariant in $M$.

### 11.0.1    *Examples*

Some examples of currents are given below:

**1.** All examples of currents (foliation) constructed from invariant measures in §2.4 give examples of foliation cycles.

**2.** If $\mu$ is a Borelian measure (positive not necessarily $\mathcal{F}$-invariant) over $M$ a current of integration $C_\mu \colon \varphi \mapsto \int_M \varphi(X_\mathcal{F}) d\mu$ is a foliation current for $\mathcal{F}$; by the above Theorem of Sullivan $C_\mu$ is a foliation cycle if, and only if, $\mu$ is $\mathcal{F}$-invariant.

**3.** Let $\mathcal{F}$ foliation transverse to the fibers of the bundle $\pi \colon E \xrightarrow{F} B$ with global holonomy $\mathrm{Hol}(\mathcal{F}) \subset \mathrm{Dif}(F)$, then given any Borelian measure $\mu_0$ in $B$ a current $C_\mu$ associate to the measure $\mu$ defined by $\mu_0$ em $E$ is a foliation current (as in 2. above) which is a foliation cycle if, and only if, $\mu_0$ and $\mathrm{Hol}(\mathcal{F})$-invariant.

**4.** Let $\mathcal{F}$ foliation of codimension 1 in $M$; according to Haefliger's theorem (Chapter 5) if $\mathcal{F}$ has a transverse closed curve homotopic to zero in $M$ then there exists leaf $L_0$ of $\mathcal{F}$ and loop (of holonomy) $\alpha_0 \in \pi_1(L_0)$ with holonomy $f_{\alpha_0} \in \mathrm{Dif}((-\varepsilon, \varepsilon), 0)$ such that $f|_{(-\varepsilon, 0]}$ is the identity and $f|_{(0, +\varepsilon)}$ is increasing.

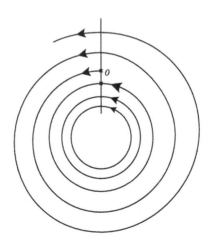

Fig. 11.1

Such a leaf $L_0$ we will call in general a *ressort leaf*, more precisely we have:

**Definition 11.2 ([Godbillon (1991)] Chapter IV, page 228).** A leaf $L_0$ of a transversely orientable codimension one foliation $\mathcal{F}$ is *ressort* (from the French word for "spring") if it contains a loop $\alpha_0$ whose corresponding holonomy map $h_{\alpha_0} \in \text{Diff}(\mathbb{R}, 0)$ is not decreasing and is, in at least of the sides of $\mathbb{R} \setminus \{0\}$, a contraction accumulating $L_0$ at itself.

Let us see how the existence of an invariant measure for $\mathcal{F}$ restricts the existence of ressort leaves.

**Proposition 11.1.** *Let $\mathcal{F}$ be a transversely orientable codimension one smooth foliation on a compact manifold $M$. The support $K$ of an invariant transverse measure $\mu$ for $\mathcal{F}$ does not contain a ressort leaf.*

**Proof.** First we observe that $K$ is invariant (because $\mu$ is invariant), closed (by definition) and any leaf contained in $K$ is dense in $K$ (indeed, $K$ is support of $\mu$). Since the measure $\mu$ is finite in compact sets we conclude that $K$ does not contain a ressort leaf. $\qquad\qquad\square$

The above proposition is enough for our purposes, *i.e.*, the homological proof of Novikov's compact leaf theorem. Nevertheless, under the hypothesis of Proposition 11.1, it is possible to say more.

**Theorem 11.1.** *Let $\mathcal{F}$ be a smooth transversely orientable codimension one foliation on a compact manifold $M$, equipped with an invariant transverse measure $\mu$ for $\mathcal{F}$. Let $K = \text{supp}(\mu) \subset M$ be the support of $\mu$. Then:*

*(1) Either $K = M$, or $K$ is a union of compact leaves and finitely many exceptional minimal sets.*

*(2) If $K = M$ then all the leaves of $\mathcal{F}$ have trivial holonomy.*

*(3) If $\mathcal{F}$ has class $C^2$ then $K$ contains no exceptional minimal set, only compact leaves.*

The proof of Theorem 11.1 requires some features from and more knowledge on the structure of codimension one foliations as Dippolito's theory on semi-stable leaves and Cantwell-Conlon's theory on minimal sets of $C^2$ codimension one foliations on closed manifolds. This is partially done in the appendix (see Appendix A). For Dippolito's theory we give the main steps, while for Cantwell-Conlon's theory we suggest their book [Candel and Conlon (2000)].

## 11.0.2     *Homological proof of Novikov's compact leaf theorem*

Note that above we strongly use the fact that $\mathcal{F}$ is of codimension 1. Suppose now that $\dim \mathcal{F} = 2$ and $\dim M = 3$ so that $\mathcal{F}$ of codimension 1. We will also assume $M$ compact and that $\mathcal{F}$ has a vanishing cycle, say, in the leaf $L_0$ of $\mathcal{F}$. We will show how to construct the foliation cycle for $\mathcal{F}$; there is no loss of generality if we assume that the vanishing cycle is *simple*: recall that (cf. Chapter 6) a vanishing cycle of $\mathcal{F}$ in the leaf $L_0$ consists of a lace (closed) $\alpha_0 \colon [0,1] \to L_0$ such that it extends to a continuous application $\alpha \colon [0,1] \times [0,1] \xrightarrow{C^0} M$ with the following properties:

(i) Given $t \in [0,1]$ the application $\alpha_t \colon [0,1] \to L_t$, $\quad \alpha_t(s) = \alpha(t,s)$ defines a loop in the leaf $L_t$ of $\mathcal{F}$.

(ii) $\alpha_0$ is the loop originally given in $L_0$.

(iii) $\alpha_0$ is not homotopic to zero in $L_0$ but $\alpha_t$ is homotopic to zero in $L_t$ $\forall t \in (0,1]$.

(iv) Fixed $s \in [0,1]$ the curve $C_s \colon [0,1] \to M$ $\quad t \mapsto \alpha_t(s)$ is transverse to the foliation $\mathcal{F}$.

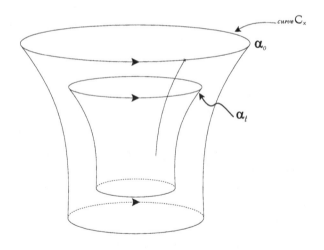

Fig. 11.2

The vanishing cycle is called *simple* when also we have

(v) the lift of $\alpha_t$, denoted by $\hat{\alpha}_t$, to the universal covering $\widetilde{L}_t$ of the leaf $L_t$ is, for each $t \neq 0$, the closed curve (because $\alpha_t \sim 0$ in $\pi_1(L_t)$) which is simple (that is, without self-intersection).

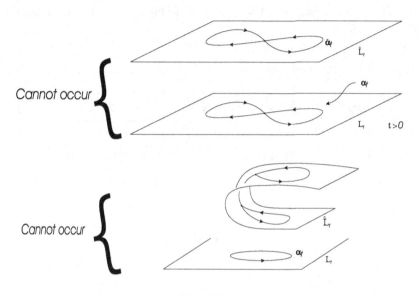

Fig. 11.3

We can approximate continuous functions by functions $C^1$ so that we can assume that $\alpha \colon S \times [0,1] \to M$ is of class $C^1$.

Note that the universal covering $\widehat{L}_t$ of $L_t$ is necessarily (diffeomorphic to) $\mathbb{R}^2$ because otherwise $\mathcal{F}$ would have some leaf covered by $S^2$, this leaf would be compact and being orientable it would be the sphere with $g \geq 0$ aisles; if $g = 1$ the universal covering is $\mathbb{R}^2$ and if $g \geq 2$ then the universal covering (as a Riemman surface) is the unit disc $\mathbb{D} \subset \mathbb{R}^2$ so we must have $g = 0$ and the leaf would be diffeomorphic to $S^2$.

By Reeb global stability theorem $\mathcal{F}$ would be a compact fibration over the circle $S^1$ with fibers $S^2$ and in this case it could not have vanishing cycle (all the leaves would be simply-connected).

Now, since each leaf $L_t$ is covered by $\widehat{L}_t \simeq \mathbb{R}^2$ each (simple) curve $\widehat{\alpha}_t$ in $\widehat{L}_t$ is boundary of a disc $\widehat{D}_t \subset \widehat{L}_t$ this allows us to obtain an immersion $C^1$, $A \colon D^2 \times (0,1] \to M$ of the solid cylinder (not compact) $D^2 \times [0,1]$ in $M$ with the following properties:

(vi)  $A_t|_{S^1 = \partial D^2} = \alpha_t$, $\quad \forall t \in (0,1]$ and the image $A_t(D^2) \subset L_t$, $\forall t \in (0,1]$.

(vii)   Given an oriented transverse flow $\vec{X} \pitchfork \mathcal{F}$ chosen from the beginning from the transverse orientability of $\mathcal{F}$ in $M$ we have which $A_t\colon D^2 \to L_t$ define, for $0 < \delta \leq t \leq 1$, lift of $A_1\colon D^2 \to L_1$ by the transverse flow $\vec{X}$

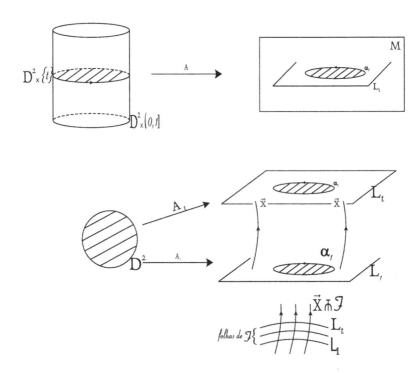

Fig. 11.4

Since $\alpha_0$ is not homotopic to zero in $L_0$ and since for each $x \in S^1$ the curve $[0,1] \to M$, $x \mapsto A(x,t)$ has a limit when $t \to 0^+$ we conclude that

(viii) The set $W = \{x \in D^2 ; t \mapsto A(x,t)$ has a limit when $t \mapsto 0^+\}$ which is an open neighborhood of $S^1$ in $D^2$ with $S^1 \subset W \subsetneq D^2$.

We denote by $C_t$, for each $0 < t \leq 1$, the foliation current of $\mathcal{F}$ defined by the integral

$$C_t(\varphi) := \frac{1}{\mathrm{vol}(D_t)} \int_{D_t} \varphi, \text{ where } \varphi \in A_c^p(M)$$

(note that still we are not using the fact that $M$ is compact, which will be used in what follows), and where $D_t = A_t(D^2) \subset L_t$ (note which $A_t\colon D^2 \to$

$D_t \subset L_t \subset M$ is a $C^1$ immersion). We obtain then, using the terminology of [Sullivan (1976)], a *family of Plante* of foliation currents $\{C_t\}_{t \in (0,1]}$ defined by the properties below which can be easily verified:

(ix) Each $C_t$ has mass 1 so that $\{C_t\}_{t \in (0,1]}$ is a pre-compact family of currents (weak topology).

(x) Each accumulation $C$ of $\{C_t\}_{t \in (0,1]}$ is necessarily a foliation cycle of $\mathcal{F}$: in fact if $t_n \searrow 0$ is such that $C_{t_n} \to C$ then the fact that the quotient $\dfrac{\text{length}(\partial D_{t_n})}{\text{Área}(D_{t_n})} \to 0$ implies that the mass of the derivative $dC_{t_n}$ satisfies mass $(dC_{t_n}) \to 0$ and therefore mass$(dC) = 0$. We obtain then foliation cycles $C$ for $\mathcal{F}$ in $M^3$, if we suppose $M^3$ compact, from the existence of a vanishing cycle for a leaf $L_0^2$ of $\mathcal{F}$, foliation of codimension 1 in $M^3$.

Such facts have been generalized by D. Sullivan for higher dimension with the notion of vanishing cycle of dimension $= \dim \mathcal{F} - 1$ (cf. [Sullivan (1976)]).

We can now conclude the following:

$$\left[ \begin{array}{c} \mathcal{F} \text{ foliation orientable and transversally orientable } C^2 \text{ of} \\ \text{codimension 1 of } M^3 \text{ compact, } \mathcal{F} \text{ with a leaf } L_0 \\ \text{containing a vanishing cycle} \end{array} \right]$$

$$\Downarrow$$

$$\left[ \mathcal{F} \text{ has a foliation cycle } C \text{ whose support contains a leaf } L_0 \right]$$

$$\Downarrow$$

$$\left[ \begin{array}{c} \mathcal{F} \text{ admits an invariant transverse measure } \mu \text{ whose support} \\ \text{contains } L_0 \text{ (in fact } \text{supp}(\mu) = \text{supp}(C)) \end{array} \right]$$

$$\Downarrow$$

since $\mathcal{F}$ is of codimension 1 and $M$ compact we have that $K = \text{supp}(\mu)$ (contains $L_0$) does not contain ressort leaf and is contained in the union of minimal sets of $\mathcal{F}$ in $M$. Thus $K$ is a union of compact leaves and hence $L_0$ is a compact leaf of $M$ that is,

$$\left[ M \text{ has } L_0 \text{ as compact leaf.} \right]$$

This ends the homological demonstration of Novikov's compact leaf theorem.

■

## Appendix A

# Structure of codimension one foliations: Dippolito's theory

In what follows, we study the structure of codimension one foliations. Our aim is to introduce the results of Dippolito and Cantwell-Conlon on semi-stability, minimal sets and structure of codimension one foliations. On this course we pave the way to prove Theorem 11.1.

### A.1  Semi-proper leaves, Dippolito's semi-stability

In what follows $\mathcal{F}$ is a codimension one transversely orientable foliation on a manifold $M$, not necessarily compact. We start with the definition of proper leaf.

**Definition A.3.** A leaf $L$ of $\mathcal{F}$ is *proper* if it is locally path-connected.

It is an exercise to show that a leaf $L$ of $\mathcal{F}$ is proper if, and only if, its topology coincides with the one induced by $M$. Or else, that a leaf $L$ is proper if, and only if, $L$ does not cluster on itself (see [Dippolito (1978), p. 408] or [Candel and Conlon (2000), Def. 4.3.3]). A compact leaf is always proper. Now we define semi-proper leaves. There are many ways of defining semi-proper leaves. We follow the nice exposition in [Seitoh (1983)] page 96. Since $\mathcal{F}$ is smooth and transversely orientable, we can choose a one-dimensional smooth foliation $\mathcal{T}$ which is everywhere transverse to $\mathcal{F}$. Denote by $\pi(\mathcal{T})\colon \widetilde{M} \to M$, the unit tangent bundle to $\mathcal{T}$. Given a leaf $L$ of $\mathcal{F}$, the *sides* of $L$ are the leaves $\tilde{L}$ of the foliation $\pi(\mathcal{T})^*(\mathcal{F})$ that project onto $L$, *i.e.*, $\pi(\mathcal{T})(\tilde{L}) = L$.

**Definition A.4.** A side $\tilde{L}$ of a leaf $L$ of $\mathcal{F}$ is *a proper side* if a transversal curve $\delta\colon [0,1] \to M$ to $L$, with starting from $L$ in the direction $\tilde{L}$, satisfies $\delta(0,\epsilon) \cap L = \emptyset$ for some $\epsilon > 0$. A leaf with a proper side is called *semi-proper*. In this case this side is called *positive side* of the leaf.

In other words, a leaf is *semi-proper* if it is proper or does not cluster on itself from one side, which is called *positive* (see [Candel and Conlon (2000), p. 118]).

The following equivalent formulation is left as an exercise to the reader:

**Lemma A.1.** *For a leaf $L$ of a codimension one transversely orientable foliation $\mathcal{F}$, the following conditions are equivalent:*

*(1) $L$ is semi-proper.*

*(2) Either $L$ is proper or it is exceptional and such that for every leaf $T$ of $\mathcal{T}$, the intersection points $L \cap T$ are the extremities gaps of $\overline{L} \cap T$, i.e., of the connected components of $T \setminus (\overline{L} \cap T)$.*

Notice that, because $\mathcal{F}$ is oriented by $\mathcal{T}$, the extremities of the gaps of $\overline{L} \cap T$ are all on a same side of $L$, the side where $L$ is semi-proper.

Now we state a notion of semi-stability:

**Definition A.5.** A semi-proper leaf $L$ of $\mathcal{F}$ is *semi-stable* on the positive side if there exists an open connected subset $U \subset M$, which is a foliated product by $\mathcal{F}$ times $\mathcal{T}$, having $L$ as a positive boundary leaf. A semi-proper leaf $L$ is *attracting* on the positive side if there is an open connected subset $U$ invariant by $\mathcal{F}$, having $L$ as positive boundary leaf, and such that $L$ is the adherence of every leaf of $\mathcal{F}$ in $U$.

The equivalent formulation found in [Candel and Conlon (2000)] is:

**Proposition A.2.** *A leaf $L$ of $\mathcal{F}$ is semi-stable if it is semi-proper and on the proper side of $L$ and in a transverse arc $J$ on this side, that meets $L$ only at a point $x_0 \in L$, the fixed points of the holonomy of $L$ cluster at $x_0$ (by definition, a point $x \in J$ is a fixed point of the holonomy if, for every loop $\gamma \subset L$ based at the point $x_0$, the corresponding holonomy map $h_\gamma$ either is not defined at the point $x \in J$ or it fixes the point $x \in J$, i.e., $h_\gamma(x) = x$, see [Candel and Conlon (2000), p. 134]).*

Now we introduce the notion of foliated product:

**Definition A.6.** Given an open $\mathcal{F}$-invariant subset $U \subset M$, where $\mathcal{F}$ is a transversely oriented codimension one smooth foliation on $M^m$, we shall say that $U$ is a *foliated product* (with respect to a one-dimensional positively oriented transverse foliation $\mathcal{T}$ to the foliation $\mathcal{F}$) if the restriction $\mathcal{T}|_U$ is a fibration by open intervals over a certain manifold of dimension $m - 1$.

Exercise A.1.1. Show that in the definition above there is a smooth diffeomorphism taking $U$ onto the product $N \times (-1, 1)$, where $N$ is a $m - 1$-dimensional manifold. Also show through an example that the foliation $\mathcal{F}$ does not need to be diffeomorphic to the horizontal product foliation on $N \times (-1, 1)$.

Let us now recall Dippolito's semi-stability theorem in [Dippolito (1978)] (see also [Candel and Conlon (2000), sections 5.2, 5.3], [Godbillon (1991), section IV. 4] and [Hector and Hirsch (1987), Chapter V, sections 3, 4]).

**Theorem A.2 (Dippolito's semi-stability).** *Let $L$ be a leaf of $\mathcal{F}$ which is semi-proper on the positive side. If the holonomy group of $L$ is not attracting on this positive side, then there exists a sequence $\{U_n\}_{n \in \mathbb{N}}$ of nested open connected subsets $U_n \subset M$, such that:*

*(1) Each $U_n$ is a foliated product of $\mathcal{F}$ times $\mathcal{T}$;*
*(2) $L$ is the leaf positive boundary of each of the open sets $U_n$;*
*(3) $\bigcap\limits_{n \in \mathbb{N}} U_n = \emptyset$*

*Therefore, if the holonomy group of $L$ is not attracting on the positive side then $L$ is semi-stable on the positive side.*

Even though it may seem natural, the well-known proof of Dippolito's semi-stability requires some deeps result, which is developed in the next section. This is also useful in the proof of Theorem A.2.

## A.2    Completion of an invariant open set

We shall present the celebrated Dippolito's structure theorem. In order to do this we need to introduce the notion of *completion* of an invariant open subset of a closed foliated manifold. We follow [Godbillon (1991)] and [Candel and Conlon (2000)]. We begin with a brief discussion of the standard theory of saturated open sets, that we use in the sequel. A more detailed exposition is found in the book of Candel and Conlon ([Candel and Conlon (2000)] Chapter 8).

We consider a codimension one, non-singular foliation $\mathcal{F}$ of class $C^\infty$ in a closed $C^\infty$ manifold $M$. Given such a pair $(M, \mathcal{F})$, we denote by $\mathcal{O}(\mathcal{F})$ the set of all open $\mathcal{F}$-saturated subsets of $M$. We assume that $\mathcal{F}$ is transversely oriented, and we let $\mathcal{L}$ be a one dimensional oriented foliation, defined by a smooth non-singular vector field transverse to $\mathcal{F}$.

Let $U \in \mathcal{O}(\mathcal{F})$ be connected; fix a Riemannian metric on $M$ and take its restriction to $U$. Let $d : U \times U \to [0, \infty)$ be the induced topological metric, and denote by $\widehat{U}$ its completion with respect to this metric. One has (see propositions 5.2.10 to 5.2.12 in [Candel and Conlon (2000)]):

**Proposition A.3. i)** *The space $\widehat{U}$ is a complete connected, $C^\infty$ manifold with finitely many boundary components, and its interior $Int\widehat{U}$ is diffeomorphic to $U$.*

**ii)** *The manifold $\widehat{U}$ has a foliation $\widehat{\mathcal{F}}$ induced from that in $U$, and the inclusion $i : U \hookrightarrow M$ extends to a $C^\infty$ immersion $\widehat{i} : \widehat{U} \hookrightarrow M$ that carries leaves of $\widehat{\mathcal{F}}$ diffeomorphically onto leaves of $\mathcal{F}$.*

**iii)** *If we let $\delta U = \widehat{i}(\partial\widehat{U})$ be the image of the boundary of $\widehat{U}$, then $\delta U$ is a union of leaves of $\mathcal{F}$, and if $L$ is a leaf in $\delta U$ then $\widehat{i}^{-1}(L)$ consists of one or two leaves in $\partial\widehat{U}$.*

**iv)** *There is also an induced oriented foliation $\widehat{\mathcal{L}}$ on $\widehat{U}$, defined by a vector field transverse to $\widehat{\mathcal{F}}$, which is carried by $\widehat{i}$ into the foliation $\mathcal{L}$.*

**Definition A.7.** The manifold $\widehat{U}$ is the (*abstract transverse*) *completion* of $U \in \mathcal{O}(\mathcal{F})$. The set $\delta(U)$ is the *border* of $U$; the leaves in $\delta U$ are the *border leaves* of $U$.

That is, a border leaf $L \subset \delta U$ is the image under $\widehat{i}$ of a leaf in the boundary of the completion $\widehat{U}$, and there are at most two such leaves in $\partial\widehat{U}$ corresponding to $L$.

A *biregular* cover of $(M, \mathcal{F}, \mathcal{L})$ is a cover $M = \bigcup U_\alpha$ by open sets $U_\alpha$ equipped with coordinates $(x_\alpha, y_\alpha) \in U_\alpha$, such that $\mathcal{F}|_{U_\alpha}$ and $\mathcal{L}|_{U_\alpha}$ are trivial, the plaques of $\mathcal{F}$ in $U_\alpha$ are the level sets of $y_\alpha$ and the plaques of $\mathcal{L}$ in $U_\alpha$ are the level sets of $x_\alpha$. Every foliated manifold $(M, \mathcal{F}, \mathcal{L})$ admits a biregular cover, which can be taken to be finite if $M$ is compact.

Recall that (cf. Definition A.6) $U \in \mathcal{O}(\mathcal{F})$ is a *foliated product* (with respect to $\mathcal{L}$) if the restriction $\mathcal{L}|_U$ fibers $U$ by open intervals over some $(m-1)$-manifold $N$. Since $\mathcal{L}$ is defined by a vector field, this bundle is trivial. Thus $U$ is diffeomorphic to $N \times (0, 1)$, but the restriction $\mathcal{F}|_U$ is not necessarily a product foliation.

One has that a leaf $L \in \mathcal{F}$ is *semi-proper* if and only if it is a border leaf of some $U \in \mathcal{O}(\mathcal{F})$ (see [Candel and Conlon (2000), Lemma 5.3.2]). Therefore each component of $\partial\widehat{U}$ is identified with a semi-proper leaf of $\mathcal{F}$; some pairs of components may be identified with a same leaf.

Using this we can restate Dippolito's semi-stability theorem as follows ([Dippolito (1978)], see also [Candel and Conlon (2000), sections 5.2, 5.3], [Godbillon (1991), section IV. 4] and [Hector and Hirsch (1987), Chapter V, sections 3, 4]).

**Theorem A.3 (Semi-stability theorem of Dippolito).** *Let $L$ be a semiproper leaf which is semi-stable on the proper side defined by the transverse arc $J = [x_0, y_0)$. Then there is a point $y_1 \in J \setminus \{x_0\}$ such that the $\mathcal{F}$-saturation $U = Sat_{\mathcal{F}}((x_0, y_1))$ is a foliated product having as border leaves the (distinct) leaves through $x_0$ and $y_1$. Also, there exists a sequence $\{y_k\}_{k=1}^{\infty} \subset (x_0, y_1]$ converging monotonically to $x_0$, such that the leaf $L_k \ni y_k$ is carried by the $\widehat{\mathcal{L}}$-fibration $\pi \colon \widehat{U} \to L$ homeomorphically onto $L$, for all $k \geq 1$.*

For the interested reader, we shall give a sketch of the proof of the above theorem. Our proof will then be based on an equivalent construction of the completion of an open invariant subset, which we pass to present.

### A.2.1   *Completion of an invariant open set - revisited*

Now we shall recall an equivalent way of introducing the completion of an invariant open subset of a foliated manifold. More precisely, we have: $M$ a closed (compact and without boundary) manifold, $\mathcal{F}$ a transversely oriented codimension one foliation of class $C^r$, $r \geq 1$. Also consider a $C^r$ one-dimensional foliation transverse to $\mathcal{F}$ and oriented according to the transverse orientation of $\mathcal{F}$. We take a finite biregular covering $\mathcal{U} = \{U_1, ..., U_s\}$ of $M$, with respect to $\mathcal{F}$ and $\mathcal{T}$. We can assume that for each $i \in \{1, ..., s\}$, there is a diffeomorphism $\xi_i \colon V_i \to \mathbb{R}^m = \mathbb{R}^{m-1} \times \mathbb{R}$, defined in some neighborhood $V_i$ of $U_i$ in $M$, such that $\xi(U_1) = (-1, +1)^m$. Moreover, this diffeomorphism takes $\mathcal{F}$ and $\mathcal{T}$ onto the horizontal and vertical foliations respectively, always preserving the orientations. The following lemma is proved in the standard manner, we leave it to the reader:

**Lemma A.2.** *If $U \subset M$ is a proper invariant open subset then $\mathcal{U}$ can be chosen such that the following properties are also true:*

*(1) The "external" plaques $\xi^{-1}(\mathbb{R}^{m-1} \times \{a\}), a = \pm 1$ are* **not** *in the boundary of $U$;*

*(2) There is $\mathbb{N} \ni r \leq s$ such that the open set $U_i$ is contained in $U$ if, and only if, $i > r$.*

*(3) For $i \leq r$ the intersection $U \cap U_i$ contains a countable number of connected components. These are denoted by $U_{ij}$, except for two of them, which adhere to the upper and lower plaques (i.e., the external plaques in the obvious sense). These two plaques are denoted by $U^+$ and $U^-$ respectively.*

Now we denote by $P_{ij}^-$ the upper plaque of $U_i$ bounding $U_{ij}$ and by $P_{ij}^+$ the lower plaque of $U_i$ bounding $U_{ij}$. Finally, we denote by $P_{ij}^-$ the upper plaque of $U_i$ bounding $U_i^-$ and by $P_i^=$ the lower one for $U_i^+$.

By gluing the manifolds $\widehat{U}_i = U_i$ for $i > r$, the manifolds $\widehat{U}_i^- := U_i^- \cup P_i^-$, $\widehat{U}_{ij} := P_{ij}^+ \cup U_{ij} \cup P_{ij}^-$ and $\widehat{U}_i^+ = P_i^+ \cup U_i^+$ for $i \leq r$, by the gluing diffeomorphisms $\xi_j^{-1} \circ \xi_i$, we obtain a $m$-dimensional manifold with boundary $\widehat{U}$, of class $C^r$, equipped with a codimension one foliation $\widehat{\mathcal{F}}$, transversely oriented, with a 1-dimensional transverse foliation $\widehat{\mathcal{T}}$, of class $C^r$. The result is:

**Proposition A.4.** *The above construction gives a manifold $\widehat{U}$ and foliations $\widehat{\mathcal{F}}, \widehat{\mathcal{T}}$, which are homeomorphic to the completion of the invariant open subset $U$ by $\mathcal{F}$ and $\mathcal{T}$, defined by a Riemannian metric on $M$, as introduced above.*

We denote by $\partial\widehat{U}$ the boundary of $\widehat{U}$. Notice that the lifted foliation $\widehat{\mathcal{F}}$ is tangent to this boundary. Therefore, the foliation $\widehat{\mathcal{T}}$ is transverse to $\partial\widehat{U}$. We can put $\partial^+\widehat{U}$ for the subset of the boundary where the orbits of $\widehat{\mathcal{T}}$ points inwards to $\widehat{U}$ and by $\partial^-\widehat{U}$ the other part.

We end this paragraph with the statement of Dippolito's structure theorem:

**Theorem A.4.** *Let $\mathcal{F}$ be a codimension one foliation on $M$, $U \in \mathcal{O}(\mathcal{F})$ a proper subset, with completion $\widehat{U}$, equipped with the foliation $\widehat{\mathcal{F}}$ and transverse foliation $\widehat{\mathcal{T}}$. Then, there are $m$-dimensional submanifolds with corners $K$ and $B$ of $\widehat{U}$ (one of them may be empty), invariant by $\widehat{\mathcal{T}}$ such that:*

*(1) $\widehat{U} = K \cup B$.*

*(2) $K$ is compact and connected while $B$ has no compact connected component.*

*(3) $C = K \cap \partial\widehat{U}$ is a compact $m - 1$-dimensional subvariety and it is the saturated of the boundary $\partial C$ by $\widehat{\mathcal{T}}$.*

*(4) $B$ has only finitely many connected components, each one having as boundary a connected component of $D$.*

*(5) The intersection $A = B \cap \partial^+(\widehat{U})$ is a $m - 1$-dimensional subvariety of the boundary $\partial\widehat{U}$ and the orbits of $\widehat{\mathcal{T}}$ define a fibration by compact intervals of $B$ onto $A$.*

**Proof.** Let us use the notation introduced in the second construction of the completion of an invariant open subset above. We denote by $V \subset \widehat{U}$ the union of open subsets $\widehat{U}_{ij}$ where $i \leq r$. Then $V \subset \widehat{U}$ is an open, $\widehat{\mathcal{T}}$-invariant subset. Moreover, it is easy to see that given the set $\widehat{U}_{ij}$, its projections onto the external plaques $P_{ij}^+$ along the orbits of $\widehat{\mathcal{T}}$, define a fibration by compact intervals, of $V$ onto the intersection $V \cap \partial^+\widehat{U}$. Since $\widehat{U} \setminus V$ is compact, we can choose a compact submanifold with boundary $S \subset \partial^+\widehat{U}$, of dimension $m - 1$, such that:

(1) $\partial^+\widehat{U} - S$ has a finite number of connected components, each of these corresponds to a connected component of the boundary $\partial S$. Each such component is non-compact. They are called *ends*.
(2) The intersection of $S$ with each connected component of $\partial^+\widehat{U}$ is a neighborhood of the intersection $(\widehat{U} - S) \cap \partial^+\widehat{U}$.

With this, we define $A$ as the adherence of $\partial^+\widehat{U} - S$ and $B$ as the part of $V$ above $A$. We also get $K$ as the adherence of $\widehat{U} - B$, if we choose $S$ such that $K$ is connected.

$\square$

We call $K$ above as the *core* of $\widehat{U}$ while the connected components of $B$ are the *arms* of $\widehat{U}$ defined by the core $K$.

Using this notion we can redefine an invariant connected open subset $U \in \mathcal{O}(\mathcal{F})$ as a *foliated product* (by $\mathcal{F}$ and $\mathcal{T}$) if its completion $\widehat{U}$ is fibred in compact intervals by the orbits of $\widehat{\mathcal{T}}$.

Then we have:

**Lemma A.3.** *Given a connected open subset $U \subset M$ which is a foliated product for $\mathcal{F}$ and $\mathcal{T}$ and its boundary leaves $L_1, L_2$, we have that:*

- *The limit sets of $L_1$ and $L_2$ coincide and are contained in the limit set of any leaf $L$ of $\mathcal{F}$ in $U$.*
- *If one of this leaves is exceptional then the same holds for the other and we have $\overline{L_1} = \overline{L_2} \subset \overline{L}$.*

The proof is quite standard and we leave it to the reader.

## A.3     Proof of Dippolito's semi-stability theorem

Now we are in conditions to prove Dippolito's semi-stability theorem.

**Proof of Theorem A.2.** We preserve the notation of the preceding sections. Given a point $p \in L$ we denote by $T^+(p)$ the positive semi-orbit of the transverse foliation $\mathcal{T}$, starting at $p$. It is easily verified that, the saturation of an open arc $J \subset T^+(p)$, with origin at $p$, and not intersecting $L$, is a connected open set $U \subset M$, having the leaf $L$ as a positive boundary leaf. Now we use the Structure theorem (Theorem A.4). Denote by $\widehat{U}$ the completion of $U$ and consider the intersection $K(L) := K \cap \widehat{L}$, where $K$ is the core of $\widehat{U}$, and $\widehat{L}$ is the leaf of $\partial^+\widehat{U}$ that corresponds to $L$. Then $K(L) \subset \widehat{L}$ is a compact, connected, $m - 1$-dimensional submanifold with boundary. We may also assume that $K(L)$ is not empty, indeed that it contains the base point $\hat{p}$ of $\widehat{L}$, above $p$.

Now a few considerations more and we are done. Denote by $H :=$ $\mathrm{Hol}(\mathcal{F}, L, p)$ the holonomy group of $L$ based at $p$, with transverse section given by the orbit $T(p)$ of $\mathcal{T}$, through $p$. If $H$ is not attracting from the positive side, then there is a monotonous sequence $(p_n)_{n \in \mathbb{N}}$ of points $p_n \in J$, converging to $p$, such that each $p_n$ is fixed by all the elements of the group $H$ which are defined at $p_n$ (this means that the element admits a representative which is defined in an interval contained in $J$, that contains $p$ and $p_n$). Because $K(L)$ is compact, for each $n$ big enough, we can lift $K(L)$ to a submanifold with boundary $K_n$ in the leaf $\widehat{L_{p_n}} \ni p_n$, that projects diffeomorphically onto $K(L)$, along the orbits of $\widehat{\mathcal{T}}$. The intersection of $K_n$ with each arm of $\widehat{U}$ is connected. Therefore, we can also lift to $\widehat{L_{p_n}}$ each of the submanifolds of $\widehat{L}$ corresponding to the basis of these arms. This way we obtain by means of projection along the orbits of $\widehat{\mathcal{T}}$, a global diffeomorphism of $\widehat{L}_n$ onto $\widehat{L}$. Therefore the open subset $U_n \subset U$ bounded by the leaves $L_n$ and $L$, which corresponds to the saturation of the subarc $J_n \subset J$ between $p$ and $p_n$, is a foliated product with respect to $\mathcal{F}$ and $\mathcal{T}$. The proof is finished.                                                      □

## A.4     Guided exercises: Cantwell-Conlon's theory

In what follows we give a sketched proof of Theorem 11.1.

**Exercise A.4.1.** In this exercise we shall prove the following result due to Cantwell-Conlon (see [Candel and Conlon (2000)] Chapter 8, pp. 190-192):

**Theorem A.5 (Cantwell-Conlon [Cantwell and Conlon (1981)]).**
*Let $\mathcal{F}$ be a codimension one smooth foliation of a connected manifold $M$ and $U \subset M$ an invariant connected open subset. Then we have:*

*(1) The union of minimal sets of the restriction $\mathcal{F}|_U$ is closed in $U$ and contains only finitely many exceptional minimal sets of $\mathcal{F}|_U$;*

*(2) The union of closed leaves of $\mathcal{F}|_U$ is closed in $U$;*

*(3) Given a leaf $L$ of $\mathcal{F}$ in $U$, the closure $\overline{L}$ contains only a finite number of minimal sets of $\mathcal{F}|_U$.*

In order to prove this result we introduce a pre-order in the leaf space $M/\mathcal{F}$ of a codimension one foliation $\mathcal{F}$ on a manifold $M$. Given leaves $L_1, L_2$ of $\mathcal{F}$ we put $L_1 \leq L_2$ if the closures are related by $\overline{L_1} \subseteq \overline{L_2}$. If we consider the class of a leaf $L$ as the set $[L]$ of all leaves having the same closure than $L$ then we have in the quotient space an order relation. We can also write $L_1 < L_2$ to denote that $\overline{L_1} \subsetneq \overline{L_2}$.

Recall that we are dealing with codimension one foliations:

**Lemma A.4.** *The leaves in the class $[L]$ of a leaf $L$, have all the same nature regarding $\mathcal{F}$. More precisely, the leaves in $[L]$ are either all proper, or all locally dense, or all exceptional.*

We define the *inferior structure* of a leaf $L$ as the union $\inf(L)$ of all leaves $L'$ such that $L' < L$. The *superior structure* $\sup(L)$ is defined in a natural analogous way.

The following is left as an exercise.

**Lemma A.5.** *For a leaf $L$ we have $\overline{L} = [L] \cup \inf(L)$ as a disjoint union. The superior structure $\sup(L)$ is an open invariant subset of $M$.*

The next result states an isolation property for exceptional minimal sets.

**Proposition A.5.** *If a leaf $L$ is exceptional such that the class $[L]$ is a minimal set, then $[L] \cup \sup(L)$ is an open subset of $M$.*

**Proof.** We leave the details to the reader. We will show that if $\{L_n\}_{n \in \mathbb{N}}$ is a sequence of leaves in the complement $L_n \subset M - \overline{L}$ and converging to $L$ then, for $n$ big enough, we have $L_n \subset \sup(L)$. Notice that such sequences exist provided that $L$ is nowhere dense in $M$. Denote by $U$ the connected component of the open subset $M - \inf(L)$ that contains $L$. We can assume that the leaves $L_n$ are contained in $U$. The class of the leaf $L$ is denoted by $[L]$. Let us denote by $W_n$ the connected component of the open subset $U - [L]$, containing the leaf $L_n$. Then $W_n$ has a boundary leaf say $L_n^0$ in the class $[L]$.

Now we have distinct cases to consider:

1. If one of the open sets $W_n$, say $W_{n_o}$ contains infinitely many leaves $L_n^0$, these leaves converge to $L_{n_o}^0$ and for $n$ big enough are all contained in $\text{sup}(L)$ (it is enough to choose a sub-sequence of leaves such that the leaves in this subsequence intersect a same arm of a Dippolito's decomposition (cf. Theorem A.4) of the completion $\widehat{W_{n_o}}$ of $W_{n_o}$, whose saturated is contained in $\text{sup}(L)$).

2. If one of the open sets say $W_{n_1}$ is a foliated product then we have $\overline{L} = \overline{L^0}_{n_1} \subset \overline{L}_{n_1}$ (this is the content of Lemma A.3).

Finally, we finish by observing that the open set $U - [L]$ has only a finite number of connected components which are not foliated products.        □

**Proof of Theorem A.5.** Item (1) is a direct consequence of the fact that $\text{sup}(L)$ is an open invariant subset of $M$. Dippolito's semi-stability theorem A.2 then shows that the exceptional minimal sets are isolated implying (2). The other properties then follow.

□

Exercise A.4.2. Finally we are in conditions to prove Theorem 11.1, whose statement we repeat here:

**Theorem 11.4** *Let $\mathcal{F}$ be a smooth transversely orientable codimension one foliation on a compact manifold $M$, equipped with an invariant transverse measure $\mu$ for $\mathcal{F}$. Let $K = \text{supp}(\mu) \subset M$ be the support of $\mu$. Then:*

*(1) Either $K = M$, or $K$ is a union of compact leaves and finitely many exceptional minimal sets.*

*(2) If $K = M$ then all the leaves of $\mathcal{F}$ have trivial holonomy.*

*(3) If $\mathcal{F}$ has class $C^2$ then $K$ contains no exceptional minimal set, only compact leaves.*

**Proof of Theorem 11.1.** In order to prove the first part we state:

**Claim A.1.** *A leaf $L_0 \subset K$ cannot accumulate from a proper side, at a semi-proper leaf $L_1$ which is attracting on this side.*

**Proof.** The claim is a consequence of Dippolito's semi-stability result (Theorem A.2) and of the following statement, whose proof we leave to the reader:

*Given a locally compact topological space $\mathcal{X}$ and $\Gamma$ a pseudogroup of local homeomorphisms of $\mathcal{X}$ preserving a locally finite measure $\mu$ we have: a point $x \in K := \operatorname{supp}(\mu)$ which is fixed by a contraction $\gamma \colon A \to B$ of $\Gamma$, has a discrete and closed orbit $\Gamma(x)$ by $\Gamma$ in $\mathcal{X}$.*

We recall that $\gamma \colon A \to B$ is a *contraction* at the fixed point $x$ if $\gamma(A) = B \subset A$ and $\bigcap\limits_{n \in \mathbb{N}} \gamma^n(A) = \{x\}$.

$\square$

Moreover, the support cannot contain either a leaf $L_0$ that accumulates at a semi-proper leaf $L_1$ from the non proper side.

Now we recall that an exceptional minimal set is contained in the adherence of every leaf which is close enough (Proposition A.5). This shows that if $K \neq M$ then $K$ is contained in a union of compact leaves and finitely many exceptional minimal sets (Theorem A.5). This ends the proof of Theorem 11.1.

$\square$

# Bibliography

R. Bott, W. L. Tu, *Differential forms in algebraic topology*. Graduate Texts in Mathematics, 82. Springer-Verlag, New York-Berlin, 1982.

M. Brunella; *A Global Stability Theorem for transversely holomorphic foliations*; Annals of Global Analysis and Geometry 15, 179-186, 1997.

M. Brunella; *Courbes entières et feuilletages holomorphes*, Enseign. Math. 45 (1999), 195-216.

C. Camacho, A. Lins Neto; *Geometric theory of foliations*, Translated from the Portuguese by Sue E. Goodman. Birkhauser Boston, Inc., Boston, MA (1985).

Candel, A., Conlon, L., *Foliations. I*. Graduate Studies in Mathematics, 23. American Mathematical Society, Providence, RI, 2000.

J. Cantwell and L. Conlon: *Poincaré-Bendixson theory for leaves of codimension one*, Trans. Amer. Math. Soc. 265 (1981).

P. Dippolito: *Codimension one foliations of closed manifolds*, Ann. of Math. **107** (1978), 403–453. MR0515731, Zbl 0418.57012.

J.-P. Demailly; *Variétés hyperboliques et équations différentielles algébriques*, Gaz. Math. No. 73 (1997), 3–23.

de Melo, W., Palis, J., *Geometric theory of dynamical systems. An introduction.*, Translated from the Portuguese by A. K. Manning. Springer-Verlag, New York-Berlin (1982).

do Carmo, M., P., *Riemannian geometry*, Translated from the second Portuguese edition by Francis Flaherty. Mathematics: Theory & Applications. Birkhauser Boston, Inc., Boston, MA (1992).

G. De Rham; *Variétés Différentiables*, Paris: Hermann, 1960.

G. De Rham; *Varietés différentiables. Formes, courantes, formes harmoniques.* Paris: Hermann 1955.

R. Edwards, K. Millet, D. Sullivan; *Foliations with all leaves compact*; Topology 16, (1977), 13-32.

D.B.A. Epstein; *Foliations with all leaves compact*; Ann. Inst. Fourier, Grenoble 26, 1 (1976), 265-282.

C. Ehresmann; *Sur les sections d'un champ d'éléments of contact dans une variété différentiable*. C. R. Acad. Sci. Paris 224, (1947). 444–445.

C. Ehresmann, G. Reeb; *Sur les champs d'éléments of contact*, C.R. Acad. Sci. Paris, 218, 1944.

E. Ghys; *Holomorphic Anosov Systems*, Inv. Math. 119, 585-614 (1995).

C. Godbillon; *Feuilletages: Études Geométriques*, Birkhäuser, Berlin 1991.

P. Griffiths, J. Harris; *Principles of Algebraic Geometry*, John Wiley & Sons, New-York, 1978.

R.C. Gunning; *Introduction to holomorphic functions of several variables, vol. I*, Wadsworth & Brooks/Cole, California 1990.

A. Haefliger; *Structures feuilletées et cohomologie à valeur dans un faisceau of groupoïdes.* Comment. Math. Helv. 32 1958 248–329.

A. Haefliger: *Travaux de Novikov sur les feuilletages*, Séminaire Bourbaki 20e année, 1967-68, Num. 339, p. 433-444, 1967-68.

A. Haefliger; *Some remarks on foliations with minimal leaves.* J. Differential Geom. 15 (1980), in the. 2, 269–284 (1981).

V. L. Hansen; *Braids and Coverings.* London Mathematical Society Student Texts (No. 18), Cambridge University Press, 1989.

Hector, G. and Hirsch, U.; *Introduction to the geometry of foliations. Part B. Foliations of codimension one.*, Second edition. Aspects of Mathematics, E3. Friedr. Vieweg & Sohn, Braunschweig (1987).

Herstein, I., N., *Topics in algebra*, Second edition. Xerox College Publishing, Lexington, Mass.-Toronto, Ont. (1975).

M. Hirsch; *Stability of compact leaves of foliations.* Dynamical systems (Proc. Sympos., Univ. Bahia, Salvador, 1971), pp. 135–153. Academic Press, New York, 1973.

M. Hirsch, S. Smale, *Differential Equations, Dynamical Systems and Linear Algebra*, Accademic Press, San Diego, CA, 1974.

M. Hirsch, W. Thurston; *Foliated bundles, invariant measures and flat manifolds.* Ann. Math. (2) 101 (1975), 369–390.

H. Holmann; *On the stability of holomorphic foliatons*; Springer Lect. Notes in Math. 198 (1980), 192-202.

H. Hopf, Ubër die Abbildungen der dreidimensionalen Sphäre auf die Kugelfläche, Math. Ann. 104 (1931), 637-665.

E. Lima; *Commuting vector fiels on $S^3$*; Ann. of Math., vol. 81, 1965, p. 70-81.

David W. Lyons; *An Elementary Introduction to the Hopf Fibration*, Mathematics Magazine Vol. 76, No. 2 (Apr., 2003), pp. 87-98.

M. McQuillan; *Diophantine approximation and foliations*, Publ. Math. I.H.E.S. 87 (1998), pp. 121-174.

M. McQuillan; *Non-commutative Mori Theory*, Prepublication I.H.E.S./M/01/42, August 2001.

J. Milnor: Morse Theory; "Annals of Mathematics Studies", Princeton University Press, 1963.

J. Milnor; *Growth of finitely generated solvable groups.* J. Differential Geometry, 2, 1968, 447–449.

R. Moussu, F. Pelletier; *Sur le théorème of Poincaré-Bendixson.* Ann. Inst. Fourier (Grenoble) 24 (1974), in the. 1, 131–148.

P. Painlevé; *Leçons sur la théorie analytique des équations différentielles*, Paris,

Librairie Scientifique A. Hermann, 1897.

J. Plante;  *Anosov Flows*,  Am. Journal of Mathematics, 94 (1972), pp. 729-754.

J. Plante;  *A Poincaré-Bendixson Theorem*,  Topology 12, pp. 177-181, 1973.

J. Plante; *On the existence of exceptional minimal sets in foliations of codimension one*. J. Differential Equations 15 (1974), 178–194.

J. Plante; *Measure preserving pseudogroups and a theorem of Sacksteder*. Ann. Inst. Fourier (Grenoble) 25 (1975), in the. 1, vii, 237–249.

J. Plante;  *Foliations with measure preserving holonomy*;  Ann. of Math., 102, 1975, pp. 327-361.

J. Plante, W. Thurston; *Polynomial growth in holonomy groups of foliations*. Comment. Math. Helv. 51 (1976), in the. 4, 567–584.

G.  Reeb;  *Sur les variétés intégrales des champs d'éléments of contact complètement intégrables*. C. R. Acad. Sci. Paris 220, (1945). 236–237.

G.  Reeb; *Variétés feuilletées, feuilles voisines*. C. R. Acad. Sci. Paris 224, (1947). 1613–1614.

G.  Reeb; *Remarque sur les variétés feuilletées contenant une feuille compacte to the groupe of Poincaré fini*. C. R. Acad. Sci. Paris 226, (1948). 1337–1339

G.  Reeb; *Sur certaines propriétés topologiques des variétés feuilletées*; Actualités Sci. Ind., Hermann, Paris, 1952.

H.  Rosenberg, R. Roussarie, D. Weil; *A classification of closed orientable 3-manifolds of rank two*, Ann. of Math. 91, no. 2, (1970), 449-464.

D. Ruelle; Statistical Mechanics,  New York, Benjamin 1969.

D. Ruelle, D. Sullivan; *Currents, Flows and Diffeomorphisms*,  Topology, vol. 14, $n^{the}$ 4, 1975.

H. Rummler; Comment. Math. Helv., 54, 1979.

P. Molino; *Riemannian Foliations*; Progress in Mathematics, Birkhäuser, Boston 1988.

G. Reeb, P. Schweitzer; *Un théorème de Thurston établi au moyen de l'analyse non standard*. (French) Differential topology, foliations and Gelfand-Fuks cohomology (Proc. Sympos., Pontifícia Univ. Católica, Rio de Janeiro, 1976), p. 138. Lecture Notes in Math., Vol. 652, Springer, Berlin, 1978.

Rotman, J., *An introduction to algebraic topology*. Graduate Texts in Mathematics, 119. Springer-Verlag, New York, 1988.

R. Sacksteder; *Foliations and pseudogroups*. Amer. J. Math. 87, 1965, 79–102.

R. Sacksteder; *On the existence of exceptional leaves in foliations of codimension one*; Ann. Inst. Fourier, Grenoble, 14, 2, 1969, pp. 224-225.

B. Scárdua, Transversely affine and transversely projective foliations, *Ann. Sc. École Norm. Sup.*, $4^e$ série, t.**30**, 1997, p.169-204

H. Seifert, *Closed integral curves in 3-space and isotopic two-dimensional deformations*, Proc. Amer. Math. Soc. 1, (1950). 287–302.

L. Schwartz; *Théorie des distributions*,  Nouvelle Édition, Paris, Hermann, 1966.

S. Schwartzmann; *Asymptotic cycles*,  Ann. Math., 66, pp. 270-284, (1957)

P. Scott; *The geometries of 3-manifolds*, Bull. London Math. Soc. 15 (1983), 401-487.

W. Schachermayer; *Addendum: Une modification standard de la démonstration non standard de Reeb et Schweitzer "Un théorème de Thurston établi au moyen de l'analyse non standard"*, Differential topology, foliations and Gelfand-Fuks cohomology (Proc. Sympos., Pontifícia Univ. Católica, Rio de Janeiro, 1976), pp. 139–140. Lecture Notes in Math., Vol. 652, Springer, Berlin, 1978.

P. A. Schweitzer, *Counterexamples to the Seifert conjecture and opening closed leaves of foliations*, Ann. of Math. (2) 100 (1974), 386–400.

Akira Seitoh; *Remarks on Stability for Semiproper Exceptional Leaves*, Tokyo J. Math. VOL. 6, No. 1, 1983.

Y. Siu, Techniques of Extension of Analytic Object, Lecture Notes in Pure and Appl. Math. 8, Marcel Dekker, Inc., New York, 1974.

D. Sullivan; *Cycles for the Dynamical study of Foliated Manifolds and Complex Manifolds*, Inventiones Math., 36, pp. 225-255, (1976).

W. P. Thurston, *Anosov Flows and the Fundamental Group, Topology*, Vol. 11, pp.147-150. Pergamon Press, 1972.

W. P. Thurston; *A generalization of the Reeb stability theorem*, Topology 31, pp. 347-352, 1974.

W. Thurston; *A local construction of foliations for three-manifolds*, Proc. Symposia in Pure Math. 27 (1975), A.M.S., 315-319.

W. Thurston; *Existence of Codimension-One Foliations*, Annals of Mathematics Second Series, Vol. 104, No. 2 (Sep., 1976), pp. 249-268.

D. Tischler; *On fibering certain foliated manifolds over $S^1$*, Topology, vol. 9 (1970), pp. 153-154.

J. Tits; *Free subgroups in linear groups*, J. of Algebra, 20, 1972, pp. 232-270.

J. Wolf; *Growth of finitely generated solvable groups and curvature of Riemannian manifolds*, Journ. Diff. Geom., 2, 1968, pp. 421-446.

# Index

Printed in the United States
By Bookmasters